Space Policy in Developing Countries

This book analyzes the rationale and history of space programs in countries of the developing world.

Space was at one time the sole domain of the wealthiest developed countries. However, the last couple of decades of the twentieth century and the first decade of the twenty-first witnessed an increase in the number of countries with state-supported space programs. At this writing, no less than 25 developing states, including the rapidly emerging economic powers of Brazil (the seventh largest), China (second largest), and India (fourth largest), possess active national space programs with proven independent launch capability or concrete plans to achieve it soon.

This work places these programs within the context of international relations theory and foreign policy analysis. The author categorizes each space program into one of three tiers of development, based not only on the level of technology used, but also on how each program fits within the country's overall national security and/or development policies. The text also places these programs into an historical context, which enables the author to demonstrate the logical thread of continuity in the political rationale for space capabilities.

This book will be of interest to students of space power and politics, development studies, strategic studies, and international relations in general.

Robert C. Harding is Associate Professor and Chair of Political Science at Spring Hill College in Mobile, Alabama, USA. His principal research interests are space policy, international security, and Latin American politics. He is the author of *Military Foundations of Panamanian Politics* (Transaction), *The History of Panama* (Greenwood), and many scholarly articles in a number of refereed journals, including *Air & Space Power Journal*.

Space Policy in Developing Countries

The search for security and development on the final frontier

Robert C. Harding

Routledge
Taylor & Francis Group

LONDON AND NEW YORK

First published 2013
by Routledge

2 Park Square, Milton Park, Abingdon, Oxon OX14 4RN
711 Third Avenue, New York, NY 10017, USA

Routledge is an imprint of the Taylor & Francis Group, an informa business

First issued in paperback 2016

British Library Cataloguing in Publication Data
A catalogue record for this book is available from the British Library

Library of Congress Cataloging-in-Publication Data
Harding, Robert C.
Space policy in developing countries: the search for security and
development on the final frontier/Robert C. Harding.
   p. cm. – (Space power and politics)
   Includes bibliographical references and index.
   1. Astronautics and state–Developing countries. 2. National
security–Developing countries. I. Title.
   UG1525.D48H37 2012
   333.9'4–dc23
                                                        2012002658

ISBN: 978-0-415-53845-9 (hbk)
ISBN: 978-1-138-72940-7 (pbk)

Typeset in Baskerville
by Wearset Ltd, Boldon, Tyne and Wear

Dedicated to my children: Trevor, Erika, and Emily
May your aspirations be as boundless as the universe in which you live.

For the wise man looks into space and he knows there are no limited dimensions.

Lao Tzu

When once you have tasted flight, you will forever walk the Earth with your eyes turned skyward, for there you have been, and there you will always long to return.

Leonardo da Vinci

Man must rise above the Earth to the top of the atmosphere and beyond, for only then will he fully understand the world in which he lives.

Socrates

Contents

Tables

Preface

Since the earliest times, humans have looked toward the night sky in the hope of finding answers to the some of the greatest unknowns in life, and imagined countless dreams among the innumerable stars and galaxies in the known universe. Having been an amateur astronomer since childhood, I too wondered in awe. In fact, one of my earliest childhood memories was watching a grainy black-and-white image of Neil Armstrong setting foot on the moon, which was a transformative experience for my young, impressionable mind. Through my parents' later gift of an inexpensive telescope, I came to know the night sky, and I anxiously waited for the cold, crystal-clear winter nights of my native Indiana to reveal their celestial bounty.

Childhood flights of fancy invariably give way to the more tangible and immediate realities of existence. I later learned that while there certainly have been those who have romantically sought to make possible humanity's journey into space simply because "it's there," the simple fact is that most of the policies that have driven modern space programs have emanated from a much more complex yet primordial impulse—the improvement of and even the survival of the state. As a result, the expectant tone of Armstrong's "one small step" speech has been forever replaced in my mind by Carl Sagan's prescient admonition that governments do not spend vast sums for pure science, and that there must be a real political purpose. This, in part, is the underlying motivator of this book: to seek to understand and to put in perspective the political, economic, and cultural rationales used by developing countries that pursue space programs.

More broadly, the question of what role space policy plays in determining a country's overall power and influence and the trajectory of its socio-economic development is more than just an academic exercise. It is an essential part of understanding the changing dynamic of the modern international system. For much of the Cold War, the United States and the Soviet Union used their respective space programs to further their foreign policies, partly as trophies to display through the course of the ceaseless ideological struggle of the time. Having been shocked into action by the

Soviet Union's launch of *Sputnik* in 1957, the United States thereafter became the undisputed leader in practically all areas of space-related technology and accomplishments. But the waning days of the Cold War and the fall of the Soviet Union brought a reorientation of policies, with accompanying widespread reductions in many areas of defense and science as part of the so-called peace dividend. Contemporaneously, the tragic losses of the shuttles *Challenger* in 1986 and *Columbia* in 2003 seemed to irrevocably taint Americans' enthusiasm for space and to slow American leadership in space activities.

Despite the massive reallocation of funds toward national security resulting from the attacks of September 11, 2001, the first decade of the twenty-first century saw a promising and inspiring rejuvenation of the United States' space program. The Bush administration proposed in 2004 that the United States would return to the moon and go beyond. However, since the throwing down of that gauntlet, the Obama administration cancelled the Constellation program in October 2010, effectively dooming any possibility of a manned mission to the moon or anywhere else outside of low Earth orbit in the foreseeable future. Given this, coupled with the retirement of the US space shuttle fleet in 2011 and the flagging of the Ares heavy-lift launcher program, American political and technological leadership of space flight is now very much in question.

But as the old saying observes, nature abhors a vacuum, and many other space actors are set to rush into the void. Besides a very capable Russia (on whom the United States now depends for human space flight) and the multinational European Space Agency, there is a growing club of developing states for whom space policy has become a means to an end, if not a national priority. Despite greater poverty indices, more frequent social and political instability, and both perceived and real ethnic divides, much of the growth in space activities in the twenty-first century will emanate from the developing world.

The abilities of rising space actors vary considerably: from the Chinese juggernaut that has in a generation literally gone from empty rice bowls to launching satellites and manned orbital missions, down to a small Zimbabwean software company that writes satellite programming language. Between these two extremes is found a vibrant and motivated group of states that, along with the already-established space powers of Europe, Japan, and Russia, collectively raises a significant question about the continued US dominance of space. In addition, the expansion of space policy in the developing world provides an interesting window into the changing nature of the state system since the Cold War, which, as Fareed Zakaria and many others have argued, will be an international system no longer defined solely by the hegemonic presence of the United States and will exhibit multipolarity in various forms. This democratization of the international system will include space activities, increasingly undertaken by the countries of the developing world.

Acknowledgments

There are many people who directly or indirectly made this book possible. Though it is impossible to name them all, they know who they are and I thank them. Specifically, I wish to thank my wife, Dora, who picked up the slack during my long bouts of research and writing. I also wish to acknowledge Amy Gorelick of the University of Florida for suggesting the project; Daniel R. Mortensen of the Air Force Research Institute for providing access to invaluable materials at the Fairchild Information Research Center, Maxwell Air Force Base, Alabama; Roger Handberg of the University of Central Florida and Agnieszka Lukaszczyk of the Secure World Foundation for their very constructive feedback on early drafts; Cliff Staten of Indiana University Southeast, for always providing a helpful ear; and to my institution, Spring Hill College, which very graciously supported my research through course releases and research grants.

Robert C. Harding
Mobile, Alabama

Introduction

Space power as national power

Earth is the cradle of mankind, but man cannot live in the cradle forever.
Konstantin E. Tsiokovsky[1]

Change in the post-Cold War period has become the standard of our time. Whether it be the changing power structure of the international system, climate change, the speed of technological innovation, or changes within our societies, the current international situation is one of constant, accelerating transformation. One area that has certainly evolved is the importance and priority given to space-related programs by a growing number of countries around the world. As the various captains of Star Trek fame have somberly declared, space really is the final frontier. But while it has been the basis for engaging science fiction, outer space nonetheless has a very down-to-Earth feature—it has become the ultimate venue for the growth of national power and socioeconomic development among a number of the world's emergent states.

This new paradigm of international relations has been evolving for over 50 years. From the Soviet Union's launch of *Sputnik* in 1957, many states began to include space-based security concerns in their foreign policies, which forced them to consider what the then-new operations in space meant for national security; they also began to integrate space-based assets into their approaches to a wide range of national development challenges, from agriculture to health improvement to the development of natural resources. Though the importance of space to national power, prestige, and potential has been less obvious in the intervening years since the heady days of the Cold War's space race, its significance has never waned and continues to increase as many states increase national space budgets. Space has, in fact, earned a permanent place at the table in matters of international conflict, peace, national and international development, and international law.

Space was at one time the sole domain of the wealthiest developed countries. The United States and the Soviet Union/Russia, and to some extent the European Union, dominated the use of space and the associated technology in the first decades after World War II. But the last couple of decades of the twentieth century and the first decade of the twenty-first witnessed an increase in the number of countries with state-supported space programs. At this writing, no fewer than 25 developing states, including the rapidly emerging economic powers of Brazil (the sixth largest), China (second largest), and India (fourth largest), possess active national space programs with proven independent launch capability or concrete plans to achieve it soon. Space programs and their related technologies are now an integral part of the strategic and developmental policies of many relatively wealthy developing states that aspire to elevate their international status, security, and economic future. A multitude of other developing states as diverse as Mexico, Nigeria, and Malaysia have established and elevated their own space policy through the creation of national space agencies and the purchase and/or production of satellites and related space technology either through state, private, or joint efforts. For these smaller and rising middle powers, the acquisition of space capabilities is now an integral component of their national policies.

Though commercial enterprise is not a focus of this study, it must be noted that as the cost of space-related technology has decreased dramatically, the expanding number of national state actors in space has been paced by the equally impressive expansion in the number of strictly commercial space companies. Communications, geospatial information, and a wide variety of other services provided by commercial satellites affect much of modern life, and also provide vital information to governments, their agencies, and business interests worldwide. This information covers many of the same areas that national governments find important to national well-being, such as weather and climate monitoring, water management, environmental observation, topographic mapping, natural disaster planning, and crop management. These services are provided commercially by a growing cadre of companies that build satellites, create the associated technologies, and are beginning to provide basic launch services, all areas that were previously the exclusive domain of state-owned space agencies.

The growth of commercial space services has been a double-edged sword for states. By 2010, the global space industry was estimated to be worth US$276.52 billion, an 18 percent increase over 2009.[2] Of this total, worldwide commercial satellite industry revenues rose 11 percent to US$160.9 billion in 2010.[3] Despite sporadic attempts to control its proliferation, commercial satellite imagery has become so good and so broadly disseminated that many national governments, for example Israel, have complained that its existence endangers national security because potential terrorists now have access to the detailed satellite imagery necessary to plan precise attacks. Until the 1990s, such high-resolution satellite imagery was almost exclusively the domain of the militaries of developed space powers, which, for national security reasons, did not generally make their

data public. And since there were a limited number of states with the capability to launch surveillance satellites, the potential sources were likewise
limited.

Those civilian satellites that did operate before the 1990s provided
imagery of a much lower spatial resolution than their military counterparts, typically not showing clear images of objects smaller than 10 meters
across. However, that situation changed with the launch of the US
company Lockheed Martin's Ikonos satellite in 1999. Its spatial resolution
of one meter meant that for the first time, no country could depend on
geographic distance and national borders to ensure state secrets. The situation became even more fluid through the 1990s and into the 2000s as the
transfer of space technology—satellites and associated technology—
became a commercially viable avenue for major satellite producers. Today,
imagery services such as Google Earth have revolutionized access to satellite imagery in the same way that cell phones have changed communications access for hundreds of millions of people around the world—they
have democratized it.

Nonetheless, the growing actual importance of space policy stands in
stark contrast to the popular perception of the significance of space in the
modern world. Indeed, more than 50 years after the launch of *Sputnik*, the
exploration of near space via the moon-landings, and various robotic missions to the solar system's planets, surveys have shown that few people in
the West still consider space as anything novel. The popular mindset has
moved on to the wonders of the "information age" and the benefits (or
detriments) of globalization. The generations of technology spawned by
those earlier days of space exploration have been indispensable in the creation of our high-tech, instantaneous world, but space and its benefits are
now so integrated into our daily infrastructure that most people do not
give it a second thought. The reactions to the *Challenger* and *Columbia*
space shuttle tragedies aside, public complacency toward the importance
of space has become the rule, rather than the exception.

Despite these popular sentiments, the recent expansion of space programs in the developing world demonstrates that national governments
have never altered their view of the importance of space for achieving and
expanding national power—militarily or socioeconomically. This expansion of space programs is especially noteworthy because it reflects an
emergent democratization of space, which is one of the most important
factors in the changing distribution of power in the current international
arena. Many countries now use satellites for communications and obtaining weather data, through ownership or simply purchase of the data. In
fact, this broadening and expansion of the usage of space and the attendant transformation of power distribution is seen by some observers as
leading to a new space race, albeit one that has yet to gain the high profile
that the previous contest had during the Cold War. This competition is
emerging as the catalyst for a new generation of space-related policies and

innovations in both established and emerging space-faring countries. Consider how one recent space-related event affected the dynamic of interstate relations.

In January 2007, the news that China had successfully tested an anti-satellite ballistic missile sent shockwaves around the world's foreign policy community. By shooting down one of its own aging satellites from low Earth orbit, China—a country that only a generation before was seen as poor by most measures—demonstrated its intent to join the existing space powers, thus attracting attention, if not commanding respect as a potential world power. China plans to land a nuclear-powered unmanned rover on the moon by 2013, and to have in place an orbital military space station later in the second decade of this century.[4]

But while China's space policy is more ambitious and better funded than those of other developing states, it is by no means unique. The next year of this twenty-first century space race saw India following up on the Chinese success by launching its own successful probe to the moon. Around the world, increasing numbers of developing countries are investing in space-related technologies, seeking partners for space projects, and even constructing launch facilities that may one day rival the established space powers of the United States, Russia, the European Union, and more recently Japan.

But what motivates a developing country, which by definition is relatively poor, to spend the comparatively large amounts of money required for these space adventures? The short answer is that, like the United States and the Soviet Union before them, developing countries pursue active space policies because of the recognition that space is, in many ways, the ultimate measure of national power, international prestige, and demonstrated national potential. Moreover, space-based assets allow states to more fully utilize their national resources and to expand the reach of domestic socioeconomic programs into areas as diverse as agriculture, education, medicine, and economic development. Thus a space program figures as an integral facet of any capable state's national security and developmental policies. The benefits of a successful space program include advanced communications, a platform for technology improvement, greatly enhanced geographic information, and, for some, expanded defensive and intelligence capabilities. Equally important, space programs can provide the host state with increased international prestige, which accrues both domestic and international advantages. Hence, developing countries are merely being rational state actors and following the path pioneered by those space-faring states that preceded them.

The practical value of space

The academic study of space and of the reasons why states have pursued space programs has been an evolutionary, and occasionally patchy, endeavor. Since the creation of the V-2, the first operational ballistic missile introduced by Nazi Germany in the waning days of World War II, policymakers and scholars alike have been interested in the development of missiles because of their ability to project military power much farther than previously possible. The advent of nuclear weapons, the growth of missile programs, and the ability to reach ever-higher atmospheric levels for intelligence gathering were largely responsible for the subsequent focus on space programs as a facet of national strategic policy. Accordingly, the crossing of the celestial threshold by the German V-2 marked the advent of national space policies, which this book defines simply as the conscious and specific use of space and its attendant technologies to promote the security and socioeconomic interests of the state.

In response to the birth of the space age, scholars wrestled with the effects and implications of rocketry and the attendant space programs in the great ideological and technological struggle of the Cold War. The importance of space as a policy focus has persisted, even intensified, in the post-Cold War era, with space policy occupying a vital role in the economic and security schemata of most major world powers as well as aspiring regional powers. While the geopolitical circumstances that initially spurred space-related policy focuses have changed, the national policy goals of the United States, Russia, the European Union, and later Japan to utilize space for security, economic development, and prestige purposes have been unremitting.

Space has become irrevocably ingrained into the fabric of national interests and the public psyche. The launches of commercial, surveillance, weather, and military reconnaissance satellites occur with such frequency around the world as to be non-news events, the decline in newsworthiness being inversely correlated with the expanding importance of space to states and societies. Since 1957, there have been over 25,000 payloads launched by the world's space actors, but the distribution of ownership was highly skewed.[5] From 1957 to 1991, just 23 states operated satellites, emerging at a paltry rate of 0.66 annually. But from 1991 to 2008, not only did an additional 23 states acquire satellites but the ownership rate increased 300 percent. This growth curve is forecast to continue unabated. During the 2006–2015 period, an average of 24 satellites is predicted to be launched annually from the United States alone.[6] Worldwide, the growth has been equally impressive. By 2008, annual state expenditures on national space programs had reached almost US$70 billion.[7]

Space has become a vital commercial market as well. A 2006 report by the US Congressional Budget Office affirmed that the worldwide launch capacity had reached over 33 commercial launches per year. This capacity was anticipated to double in the next decade, with new spaceports in Brazil, India, China, and Japan (and possibly others) becoming fully operational, regularized, and able to absorb an ever-greater percentage of world satellite business.[8]

But the importance of space goes beyond just more commercial satellites providing a greater number of people with television, internet, and telephone services. Space-based assets are considered by modern militaries to be indispensable "force multipliers," which enhance their capabilities through reconnaissance, weather surveillance, and even real-time imaging (i.e., a "camera in the sky"). Thus, for reasons of national security and development, space-based assets have come to be the *sine qua non* of the national defense policies of all space-capable states. For example, the total unclassified US defense-related space budget for 2007 was US$43.53 billion, which included surveillance, missile warning systems, nuclear detection, navigation, and communication satellite systems, principally under three agencies: the Department of Defense, the National Reconnaissance Office, and the National Geospatial Intelligence Office.[9] This allocation was 67 percent greater than the civilian space program administered by the National Aeronautics and Space Administration (NASA). Worldwide, expenditure by national space programs is projected to grow at almost five percent annually, much of it predicted to come from developing countries.[10] At this writing, 41 countries operate satellites, and more are being added every year. Over 600 state-owned satellites are planned for launch in the second decade of the twenty-first century, driven in part by the exceptional growth of the next generation of geopositioning (global positioning systems or GPS) and communications satellites. At least one-third of these launches are projected to come from developing countries, while only 26 percent are anticipated to be from the United States.

Besides the number of launches, the number of countries that are engaged in the use of space for national purposes has increased as well. While larger, more established space powers such as Russia and France are actively developing and expanding their space presence, emerging space actors (EMSAs) such as China, India, Japan, South Korea, and Israel are also expanding their space assets to ensure that they can leverage them for maximum commercial and national security advantages. Missile programs, space technology, satellite programs for reconnaissance and mapping, and military space systems of various types exist or are on the drawing boards of almost half of the developing countries. In addition, 36 different states have contributed almost 500 astronauts to the manned programs of the United States, Soviet Union/Russia, and most recently, China, India, Iran, and even Nigeria have stated official plans for autonomous manned programs.

The political significance of space

The importance of space to nation-states is evidenced in a multitude of ways. While Winston Churchill's 1946 identification of an "iron curtain" across Europe is commonly accepted as the metaphorical beginning of the

Cold War, it was the Soviet Union's successful launching of its *Sputnik* satellite in 1957 that heightened and gave form to the acute sense of urgency and competition between the East and the West that gave birth to national space policies. *Sputnik* put space programs firmly in the minds of geostrategists as an additional factor in assessing national power in the modern era. The ability to launch payloads into space became a question not only of sovereignty and national security, but eventually of economic necessity as well.

The long-term benefits and technological offshoots of a successful space program were many, and awareness of these played a considerable role in the struggle of the Cold War. First, the practical considerations were unavoidable, as tactical concerns took on a truly three-dimensional aspect with enhanced communications, improved mapping capacity, and the ability to spy on one's adversaries via satellites orbiting far above sovereign territory and out of reach of a state's defensive potential. Second, and equally important, space programs provided the host state with a venue to develop and improve technology within its own country, thus contributing to the sense of national security free from outside dependencies.[11]

Lastly, a successful space program with independent launch capability became an effective way for a state not only to achieve practical benefits (e.g., putting satellites in orbit), but also to bolster its national prestige at home and abroad. During the Cold War, space programs became the means by which the opposing superpowers could display the supposed superiority of their respective societies. Being the first to send probes to various solar system objects—the moon, Venus, and Mars in particular—became stepping stones for the United States and the Soviet Union along the path to establishing terrestrial hegemony. It is difficult to deny, for example, the symbolic power of having one's citizens orbiting the Earth or standing on the moon, even if the scientific returns were relatively modest. For almost two decades (1957–75), space was a primary locus of the national security struggles of the major Cold War powers.

Accordingly, efforts to put more powerful rocketry into production so as to boost ever larger payloads into space became the driving force in the space research of these superpowers, occupying a prime role in their pursuit of national security, and later also economic development. Copious amounts of literature have been published describing the "space race" between the United States and the Soviet Union during the Cold War, and the central role that space played in the conflict continues to inspire historical analyses, especially following the opening of previously secret Soviet and US archives of the period.

The utilization of space has taken a prime position in the geostrategic as well as developmental plans of all current and aspiring powers. The United States' military forces have openly declared their intent to reorient their policies to formally include space as a medium to employ all aspects of US national power.[12] In 1999, the US National Space Policy stated that "space is a medium like land, sea, and air [and] … the ability to access and utilize is a vital national interest … crucial to national security and socioeconomic well-being."[13] Similar policy statements have since emerged from Russia, China, and the European Union.

But because of the relatively short period in which developing countries have pursued space activities as a means of national security and socioeconomic development, much less scholarly research exists on the programs and policies of these growing regional and, in a few cases, potential world powers. In order to understand, for example, why in 2009 Brazil would allocate US$343 million and in 2010 India would budget US$1.25 billion for their respective space programs when 32 percent of Brazilians and 42 percent of Indians still live in poverty requires a theoretical framework—a systematic understanding of the role that space policy plays in the modern nation-state, and the extent to which such a framework resembles prior theoretical perspectives regarding state security and socioeconomic development in the international system.[14]

Besides the United States and the Soviet Union, other developed states, such as France, Britain, and later, Japan, all developed indigenous rocket programs, which evolved into space programs capable of launching a variety of satellites for both civilian and military uses (though Britain later abandoned its program). But no longer is space the sole domain of an elite few countries. Today, virtually every developed country has a stake in space or is planning to do so. Of the top 25 countries by GDP, only Australia has not yet formally established a national space program.

Accordingly, an increasing number of smaller and rising middle powers have sought to add or have added indigenous space capabilities to their list of national priorities. More often than not, this capability comes in the form of building (autonomously or in cooperation with other states) or purchasing satellites for a variety of both civilian and military uses. A few of these developing states have also created their own launch programs to independently pursue their space ambitions (Table 1.1). For the largest of these rising space powers—Brazil, China, and India—space programs contribute to what Dean Cheng has termed "comprehensive national power."[15] That is, space programs help to improve these countries' national economies by raising the level of science and technology and generating high-tech jobs, and also serve national security concerns through military security, intelligence gathering, and diplomacy.

In addition to tangible technological benefits, a space program also provides equally important intangible goods such as enhanced prestige, enabling a country to project the image of having achieved great-power status. Van Dyke (1964) argued convincingly that while the United States' Apollo program, and by extension the space programs of all the early actors, were driven by multiple motivators including military security, progress in science, and economic and social benefits, national prestige was the overarching *raison d'être* that acted as an catalyst for all the others.[16]

This analysis of the importance of prestige in early space programs still finds salience in today's burgeoning field of space actors. For these reasons, a space program has become an almost obligatory step in becoming a regional and/or world power. Developing states as diverse as Malaysia, Mexico, and Nigeria have all pursued space programs that have contributed to the development of technology for satellite telecommunications, global positioning systems (GPS), and surveillance, and have even produced home-grown astronauts.

The political motivation for the expansion of space programs is straightforward and conforms to the oldest tradition of international relations— the establishment of space-related abilities and technologies offers developing states powerful strategic options and important access to otherwise difficult-to-attain or unattainable technology, all of which sustain a state's sovereignty and stability. While space represented the high ground of the Cold War superpower conflict, it nonetheless remains a logical and essential step in every capable state's ambition to expand its influence in its region or even in the international system as well as to encourage its domestic economic and social development.

The level of involvement and sophistication in space activities varies considerably among developing countries—from the private subcontractor in Zimbabwe who writes computer software for satellite tracking, to the Argentine company Aeroterra that builds geographic information and remote-sensing satellites, to India's state-sponsored space program, which is advanced enough to have launched a probe to the moon. In these and many other examples, the space programs of developing countries serve much the same functions as they did for the Cold War superpowers: to gain prestige among nations and, more importantly, to increase the experience and capabilities of the country's space industry in order to make it as autonomous as practicable.

The increased competition resulting from the existence of more launch capabilities has driven down space technology costs across the globe by 34 percent over the past decade.[17] However, most aspiring space actors lack the independent launch capacity to put their technology into space, and dependence on others for launch facilities greatly reduces the perceived and real sovereign capabilities that can be achieved through a space program. So far, only a handful of developing countries have achieved or are on the verge of achieving this critical step. While states such as North Korea and Iran have received much press coverage for their forays into advanced rocketry and first satellite launches, a more comprehensive list includes a handful of rising regional powers (see Table 1.1). But while relatively few developing states have so far achieved the ability to independently launch satellites (or in the case of China, human space flight), those that have succeeded are staking some measure of their national security and continued economic development on space-related ventures.

Table 1.1 First satellites by state

Country	Year of first satellite in orbit	Satellite name	Launch site
Soviet Union	1957	Sputnik	Baikonur Cosmodrome, Kazakhstan
United States	1958	Explorer 1	Cape Canaveral, USA
United Kingdom	1962	Ariel 1	Woomera, Australia
Canada	1962	Alouette 1	Cape Canaveral, USA
Italy	1964	San Marco 1	Wallops Flight Facility, USA
France	1965	Astérix	Hammaguir, Algeria
Australia	1967	WRESAT	Woomera, Australia
(West) Germany	1969	Azur	Western Test Range, USA
Japan	1970	Ōsumi	Uchinoura Space Center, Japan
China	1970	Dong Fang Hong 1	Jiuquan Satellite Launch Center, China
Poland	1973	Intercosmos Kopernikus 500	Baikonur Cosmodrome, Kazakhstan
Netherlands	1974	Astronomische Nederlandse Satelliet	Vandenberg AFB, USA
Spain	1974	Intasat	Vandenberg AFB, USA
India	1975	Aryabhata	Kapustin Yar, USSR
Indonesia	1976	Palapa A1	Cape Canaveral, USA
Czechoslovakia	1978	Magion 1	Plesetsk, USSR
Bulgaria	1981	Intercosmos Bulgaria 1300	Plesetsk, USSR
Brazil	1985	Brasilsat A1	Plesetsk, USSR
Mexico	1985	Morelos 1	US Space Shuttle Discovery
Sweden	1986	Viking	Kourou, French Guiana
Israel	1988	Ofeq 1	Palmachim AFB, Israel
Luxembourg	1988	Astra 1A	Kourou, French Guiana
Argentina	1990	Lusat	Kourou, French Guiana
Pakistan	1990	Badr-1	Xichang Satellite Launch Center, China
South Korea	1992	Kitsat A	Kourou, French Guiana
Portugal	1993	PoSAT-1	Kourou, French Guiana
Thailand	1993	Thaicom	Kourou, French Guiana
Turkey	1994	Turksat 1B	Kourou, French Guiana
Chile	1995	FASat-Alfa	Plesetsk, Russia
Ukraine	1995	Sich-1	Plesetsk, Russia

Country	Satellite	Year	Launch site
Malaysia	*MEASAT*	1996	Kourou, French Guiana
Norway	*Thor 2*	1997	Cape Canaveral, USA
Philippines	*Mabuhay 1*	1997	Xichang Satellite Launch Center, China
Egypt	*Nilesat 101*	1998	Kourou, French Guiana
Singapore	*ST-1*	1998	Kourou, French Guiana
Denmark	*Ørsted*	1999	Vandenberg AFB, USA
South Africa	*SUNSAT*	1999	Vandenberg AFB, USA
Saudi Arabia	*Saudisat 1A*	2000	Baikonur Cosmodrome, Kazahkstan
United Arab Emirates	*Thuraya 1*	2000	Sea Launch (private), Pacific Ocean
Morocco	*Maroc-Tubsat*	2001	Baikonur Cosmodrome, Kazahkstan
Algeria	*Alsat 1*	2002	Plesetsk, Russia
Greece	*Hellas Sat 2*	2003	Cape Canaveral, USA
Nigeria	*Nigeriasat 1*	2003	Plesetsk, Russia
Iran	*Sina-1*	2005	Plesetsk, Russia
Belarus	*BelKA*	2006	Baikonur Cosmodrome, Kazahkstan
Kazakhstan	*KazSat 1*	2006	Baikonur Cosmodrome, Kazahkstan
Colombia	*Libertad 1*	2007	Baikonur Cosmodrome, Kazahkstan
Mauritius	*Rascom-QAF 1*	2007	Kourou, French Guiana
Venezuela	*Venesat-1*	2008	Xichang Satellite Launch Center, China
Vietnam	*VINASAT-1*	2008	Kourou, French Guiana
Switzerland	*Swisscube-1*	2009	Satish Dhawan Space Centre, India

Planned satellites

Country	Satellite	Year	Launch site
Latvia	*Venta-1*	2012	Satish Dhawan Space Centre, India
Romania	*Goliat*	2012	Kourou, French Guiana
Peru	*Chasqui 1*	2012	Russia
Azerbaijan	*AzerSat 1*	2012	Baikonur Cosmodrome, Kazahkstan
Ecuador	*NEE-01 Pegaso*	2012	Russia
Bangladesh	*unnamed*	2013	to be determined
Croatia	*unnamed*	2013–14	to be determined
Tunisia	*ERPSat01*	2013	to be determined
Uruguay	*unnamed*	2013	to be determined
Laos	*unnamed*	2013	to be determined
Moldova	*unnamed*	2013	to be determined
Turkmenistan	*unnamed*	2014	to be determined

In sum, there is an essential parallel to be drawn regarding the importance of this newest phase of reorganization and redefinition of state power. Just as the establishment of standing militaries in the eighteenth century became the cornerstone of the modern state, as recognized by Carl von Clausewitz's celebrated concept of the trinity of military, government, and the people, a space program has become in modern times the fourth pillar of any modern nation-state that aspires to better its lot. Given this imperative, the traditionally limiting socioeconomic term "developing" is not sufficient to exclude a host of new actors from the previously exclusive arena of space.

Rationale and organization of the book

For the purpose of this book's analysis, the term "developing countries" will be somewhat more broadly defined than is traditionally accepted in international relations, particularly in the sub-discipline of international political economy. While there is no single accepted definition of what constitutes a developing country, the two most common benchmarks are the Gross National Product (GNP) per capita, as reported by the World Bank, and the Human Development Index (HDI), produced by the United Nations Development Programme. In 2007, countries with a GNP per capita below US$11,116 were classified as "developing" by the World Bank.[18] Of the 177 countries included in the 2008 *Human Development Report*, 70 qualify as high-income, and the rest fall into either the middle or low-income categories.[19]

However, this book is not so much concerned with specific socioeconomic determinants nor with ascertaining a state's financial ability to sustain a space program, but seeks to understand the purpose that space programs play in the big picture of the national and foreign policies of developing countries. Thus, the use of "developing" herein reflects not so much a state's economic development, but the longevity of a country's efforts in space-related activities. Indonesia, Israel, and South Africa, each with young burgeoning space programs, all fall squarely within these parameters, though the traditional socioeconomic definitions apply very differently to each.

China's inclusion in this analysis requires special justification. China being the preeminent rising economic force of the twenty-first century, it might seem counterintuitive to think of it as a developing country. The images of the ultramodern metropolises of China's eastern coast, such as Shanghai, and the constant prognostications of China soon overtaking the US as the world's largest economy seem to belie the "developing country" classification. However, despite the country's astonishing economic and social transformation, China still remains solidly in the developing category because of the very incomplete nature of its change. Given China's GDP per capita of US$6,600, the International Monetary Fund ranks

China ninety-ninth of 181 countries, which puts it behind much of Latin America and even half of Africa. Moreover, despite its rapid and massive urbanization, almost half of China's 1.3 billion people remain rural peasants who are engaged in mostly subsistence agricultural activities, earn only a couple of hundred dollars annually, and lack many modern amenities such as running water. It is for these reasons that China's embrace of space power is completely relevant to this analysis. As a quickly developing state, China has committed an important portion of its resources to the goal of building an autonomous space program that is intended to further the socioeconomic and security goals of the country.

This book addresses three principal questions: (1) How do space programs fit into the traditional paradigms of international relations, and what are the policy priorities and decisions that have motivated developing states to divert relatively scarce resources toward space-oriented projects? (2) How does the brief history of space policy in developing countries compare to the histories of more established and wealthier space powers? (3) What role do the space programs of developing states play in their developmental and security schemata and how can these policy initiatives be understood comparatively and theoretically? What specific benefits do these aspiring space actors accrue from engaging in space activities?

Chapter 1 provides an introduction to space policy and presents the argument that the pursuit of space-related endeavors is part of a logical progression in a state's assurance of its national security and economic development. It elucidates the logical import of space programs by briefly tracing how other technological developments and pursuits have filtered down from being the prerogative of an elite few states to becoming part and parcel of a great many states' national security and economic development realities. Central to this analysis will be a discussion of the theoretical relationship between space programs and national power and development, and to this end, this chapter attempts to contribute to the ongoing development of a theoretical framework for the budding field of what has been termed "space power." This chapter will argue that states have traditionally structured national space policies in ways that are not at all unlike their terrestrial national security and development priorities—that, in a Hobbesian world of competitive states, space power serves to ensure not only the survival of the state but its prosperity. The chapter will also examine multilateral efforts to create international space regimes, conceived to foster peace, if not cooperation, in space, and the effects such regimes have had and will have on the ability of developing states to fulfill their space policy objectives.

Chapter 2 provides a concise examination of the evolution of rocket and space programs following World War II, focusing largely on the competition between the United States and the Soviet Union. The intent of this review is to explicate how space policy evolved to become an integral, even vital, component of the foreign policies and strategies of these Cold War rivals. This exercise aims to establish a clear understanding of how space has been utilized for political purposes since the dawn of the space age and how this dynamic remains largely intact today. This chapter also succinctly summarizes the growth of two subsequent "developed" space programs—those of the European Union and Japan—both of which offer interesting counterpoints in terms of the rationale of their space programs and serve as a useful example of an alternate trajectory for the future of space policy in the developing world.

Chapter 3 categorizes, examines, and describes the space-related activities of non-traditional space-faring developing states. The goal here is to understand where various groupings of developing states are in terms of their "space power" and to characterize their space-related activities according to national agendas, economic capabilities, and relations vis-à-vis other space actors, traditional and non-traditional alike. This chapter then examines specifically the evolution of the largest and most capable EMSAs—Brazil, China, and India—which this book terms "first tier" EMSAs. The chapter highlights some of the most noteworthy accomplishments and projects of these up-and-coming space actors. It also examines the ways in which their space programs enhance both national security and national development goals, such as the tangible and intangible benefits that accompany such programs, and how space activities contribute to the increasing phenomenon of South-South cooperation, which eschews the developed space programs in favor of technical cooperation with other EMSAs and is a source of growing competition for worldwide space-launch businesses.

Chapters 4 and 5 extend the analysis of EMSAs to the second and third tier space actors. These smaller but no less enthusiastic states now make up the majority of the world's space actors. These chapters explore the complex history and motivations that drive some of the world's poorer countries to pursue space-based technologies for socioeconomic development and, occasionally still, for national defense.

The Conclusion offers an analysis of how developing states' growing participation in space activities has affected, and may continue to affect, the dynamics of the international state system, particularly regarding the largest of the rising space actors, whose programs are most likely to have an impact on space politics in the near future.

1   Space power and the modern state

Non est ad astra mollis e terris via.
                                    Seneca

In his seminal 1994 book *Pale Blue Dot*, Carl Sagan compellingly argued that space exploration is not only an exciting endeavor, but also an indispensable undertaking that would ultimately ensure the continuation of the human species. Specifically, he asserted that civilization is obliged to become space-faring—not because of exploratory or romantic zeal, but for the most practical reason imaginable: staying alive ... if our long-term survival is at stake, we have a basic responsibility to our species to venture to other worlds.[1]

In less dire and more immediate terms, the logic of Sagan's argument is directly applicable to understanding the growing phenomenon of developing countries choosing to undertake what are among the most expensive projects that any state can assume—the development of space programs. Following this Saganist logic, states pursue space-related strategies and technologies to promote their socioeconomic prosperity and to ensure their sovereignty, making space programs just the latest in a long line of innovations meant to promote national security and socioeconomic development. In fact, an increasing number of scholars now suggest that the utilization, if not control, of outer space will eventually become the most important pillar of the national power of states, just as the control of the oceans and far-flung territories helped to ensure the prosperity, security, and hegemony of European states, and later the United States, in the centuries following Columbus's four voyages.[2] Jim Oberg (1999) has asserted the necessity of states' pursuit of space-related activities by emphatically arguing that space activities will become the key to humanity's long-term survival.[3]

Concurrent with any understanding of the role and import of space programs to the modern nation-state is the recognition that the politics, technology, and aspirations associated with space programs almost invariably go hand-in-hand with missile technology and, when technologically and politically feasible, nuclear ambitions, forming a symbiotic relationship. The history of the past 60 years of space-related activities has demonstrated this positive correlation and suggests a likely parallel trajectory for the most powerful and capable of the emerging space actors—Brazil, China, and India—which either have tried to develop or already possess the capacity for nuclear weapons and nuclear energy to accompany their advances in rocketry and space programs.

A theoretical framework for space power

Space, power, and politics have never been strangers. National political goals have always been the *raison d'être* for space programs, which Michael Sheehan (1999) notes "have reflected and implemented the prevailing national and international ideologies of the time."[4] Theoretical as well as practical advancements carry with them debates that challenge the status quo and push back the envelope of human knowledge and abilities, simultaneously reorganizing the established precepts of power and societies. Such was the case for Galileo Galilei, whose scientific championing of the Copernican heliocentric model of the solar system brought the wrath of the Catholic Church upon him, in part because his arguments challenged established teachings and, therefore, the power of the Church.

The concept of power—its manifestations, distribution, and use—is central to understanding a state's place, role, and potential in the international system, which for the purpose of this analysis now includes space. In general, power is understood as an actor's ability to influence other actors within the international system through coercive, attractive, cooperative, or competitive means. More specifically, Joseph Nye (2002) identifies three types of national power: military power, economic power, and soft power. Military power, typically the most readily apparent, refers to the ability to use force to achieve a goal. Economic power refers to a state's ability to influence and shape economic systems to further national goals via wealth creation and to privilege its own society. Soft power, the most recently recognized, refers to a state's ability to "obtain the outcomes it wants in world politics because other countries want to follow it, admiring its values, emulating its example, aspiring to its level of prosperity and openness."[5] Each of these conceptual areas is pertinent to the role that space policy now plays in national power and socioeconomic development.

But despite the modern importance of space, no widely accepted theory of space power has emerged that adequately explains states' usage of outer space and their space policies. Perhaps this is so because space has seemed to be an arena unlike any other, although activities within this arena are in fact so reflective of many of the characteristics associated with the current state system. It is a forum where one finds superpower competition, international cooperation, subterfuge, and economic opportunities all at the same time—just as in terrestrial international politics. Since the beginning of the space age, it has been commonly accepted by academics that what has guided, and continues to guide, countries' space policies is the oldest theory of international relations: motivated self-interest or realism.

As Carl Sagan astutely observed, "governments do not spend vast sums just for science and technology, or merely to explore. They need another purpose, and it must make real political sense."[6] The popular conception that space programs have been pursued by nation-states in the spirit of exploration and peaceful cooperation is not supported by the historical facts. Thus far, all major space-faring states have had a strong national security interest in space, even while simultaneously engaging in a number of non-security-related space activities. In the near term, the cooperative use of space, free from conflict, seems unlikely since space policy, as it has evolved to date, has been fashioned almost exclusively according to each state's security and self-interest.

This observation does not deny that there have indeed been selected projects that have revealed humanity's cooperative potential in space. For example, the Hubble Space Telescope, a US project with European assistance, has revealed our solar system, our galaxy, and the universe at a level of detail previously unattainable and inconceivable to terrestrial-bound astronomers; it has been an incredible, eye-opening example of cooperation that promotes science for science's sake. But the optics and technology that make the Hubble's astounding images possible are the by-products of technology developed for Cold War space-based surveillance.

The historical record demonstrates that when states undertake space projects, even those that are ostensibly cooperative in nature, the frequently unstated policy goal has usually been to further the political, strategic, and economic goals of the individual state, and not necessarily to promote "international cooperation" for the "good of all mankind."[7] In order to contextualize and better understand developing countries' space aspirations in the twenty-first century, it is necessary to sketch out a theoretical framework to show why space programs have figured so highly in the policy decisions of the more developed space-capable countries. This exercise will provide the basis for understanding the space policies of the developing world.

On Earth as it is in Heaven?

The space era may be divided into two discrete periods: the Cold War with its attendant superpower competition, and the current one, which has so far been divided between the quest for information in a more interconnected world and continuing traditional security and development needs. The development of a theory of space power will provide an opportunity to maximize the benefits of space for the global society.

Though it is now a cliché to say that the world changed after the Cold War, the foundational ideals of sovereignty, national power, and international security as traditionally understood in international relations are indeed in flux. The very concept of national sovereignty is being constantly tested on various fronts through the "war on terror," expanded economic interdependence, and global climate change. The post-9/11 period has shown especially dramatic effects on the ways in which geopolitical conceptions of national sovereignty have expanded beyond traditional thinking.[8]

But understanding what motivates interstate relationships is the *raison d'être* of international relations (IR) theory. From Thucydides' incisive analysis of conflict between Athens and Sparta in *The History of the Peloponnesian War* (*c*.400 BCE) to E.H. Carr's harsh scrutiny of the causes of World War II (1946) to Zbigniew Brzezinski's criticisms of US foreign policy in the Middle East (2004), international relations theorists have attempted to describe, analyze, explain, and *predict* states' actions within the international system.[9] While the immense literature of international relations theory spans a spectrum of theoretical perspectives, it all has had a common denominator, which is the terrestrial geopolitical limitation. Simply put, like pieces on a chessboard, all state interactions have been Earth-bound and two-dimensional, which defined the extension and limits of national power and interests as well as attempts at interstate cooperation within that range.

The advent of the space age, on the other hand, began to reduce (if not eliminate) these long-imposed limitations and to render murkier the established post-Westphalian boundaries of national sovereignty, which had given states international recognition of their autonomy. While there existed established concepts of national borders and later even international agreements regarding territorial waters and national airspace, the ability of a state to fly a craft over another state in outer space—outside the reach of other states' military power and the established precepts of international law—created new opportunities and dangers. The space age opened a new frontier that states could exploit, if only they possessed the financial and technological means to do so. Governments have ever since grappled with how to incorporate the realm of space into their understanding and interpretation of territoriality, international law, and national security. Thus, states' ability to utilize space, and perhaps even control it, has played a highly important role in helping to determine the nature and parameters of post-World War II interstate relations.

Perhaps because of its earlier unattainable nature, outer space had been previously treated within international relations as being as distant a subject as the science fiction that has always described it to most people. However, the post-World War II technological surge in rocketry and nuclear weaponry changed that interpretation. By 1955 the United States had developed a classified space policy (NSC 5520) that recognized the emerging role of space in science and military applications within the context of the Cold War, and proposed that the US State Department should work to "prevent an arms race from spreading into outer space" and to develop international law that would aid in this effort.[10] Much of the foundational language of the charter of the US civilian space agency, the National Aeronautics and Space Administration (NASA), is drawn from this document. To aid in the understanding of the role that space policy plays today in defining and expanding national power, particularly in developing countries, a general working theory of "space power" is needed. Such a rethinking of the paradigm of state power is not without precedent, but there are difficulties in finding consensus about its parameters.

In his treatise on the development of scientific knowledge, *The Structure of Scientific Revolutions* (1962), Thomas Kuhn argued that the trajectory in the development of a new theoretical paradigm is typically abrupt following the advent of an incommensurable event. This incongruity generates a "paradigm crisis," when there is no longer a basis for comparability between previously held notions of reality and current developments. Such circumstances consequently generate a paradigm shift that incorporates the novel development, though not without a period of competing and conflicting paradigms.[11] This is where the development of a theory of space power currently resides: possessing some solid pillars but yet still unsure of its own place in international relations theory. It is nonetheless essential that theory guide one's thinking regarding space activities, because theory clarifies these concepts and ideas, and until there is clarity on space-power-related concepts, the understanding around space competition, and perhaps cooperation, among states and their policymakers will be restricted at best.

There have been various efforts to establish a theory of space power, most of which reason that parallels can be drawn from earlier terrestrial experiences in international politics, on the principle that the history of international relations on Earth predicts a similar trajectory in space. The challenge that space poses to international relations theory is immense, but the history of the space age to date has mostly reflected the political inclinations that typified twentieth-century interstate relations. But, as Geoffrey Sloan and Colin Gray (1999) have argued, the elements of policy and strategy are timeless, and therefore not subject to the vagaries of any particular place or period in history.[12] Similarly, Kenneth Waltz's (1979) foundational work in structural realism posits that because of the permanent feature of anarchy in the international system, states will act above all in ways that ensure their security—in its various forms—in a self-help environment.

This position is not without its detractors, some of whom argue that space is a unique forum and that no precedents adequately inform our understanding of its implications on state policy.[13] Though we can certainly recognize the unique attributes of space, especially in the extreme physical and technological challenges to its use, we have no reason not to assume that the basic policy prescriptions and theoretical paradigms that have explained states' actions on Earth throughout history will, at least in part, shape a future state system that includes space as a permanent feature of international relations.

Precedents of space power

Those attempting to construct a theory of space power have often sought inspiration from the development of terrestrial strategic thought and the assumption that a country's space policy will be analogous to its historical terrestrial policies. Accordingly, in the spirit of Carl von Clausewitz, a state's space policy can be viewed as a continuation of its national policy by other means. Implicit in this paradigmatic construction is the assumption that a theory of space power will be derived from the established canon of international relations theory.

While it is possible to explore space power via a multitude of conceptual models, most of the literature produced thus far has been firmly couched in the oldest traditions of realism and liberalism. These two approaches regarding the role that space plays in a state's national policy have been contemplated for decades, and can be understood one of two ways: either space is a continuation of traditional interstate competition and security struggles, or space is a commons for all humanity where international cooperation will be essential.[14]

As will be described, thus far realism has been argued to most accurately explain the formative years of the first space-faring countries' space policies, while liberalism enlightens our understanding of some of the events of the waning years of the Cold War as well as the immediate post-Cold War period. For some scholars, liberalism's cooperative focus has been the chief determinant in explaining the development of non-state and particularly business space efforts in recent years.[15] A parallel implementation of the two paradigms is useful to comprehend the current environment, in which outer space is increasingly becoming a more crowded realm with an ever-increasing number of actors, though realism has never strayed far from being the main contender as the principal theoretical framework for understanding the current space arena.

The realist tradition in international relations makes a number of key assumptions about states' actions in the international system. As the oldest of international relations theories, realism assumes the absence of a single overarching international power. In the presence of anarchy, states will, as rational actors, pursue self-interested actions that assure their survival in a competitive international environment. To this end, states will pursue policies that allow them to accrue resources sufficient for this goal, producing an unvarying pursuit of power among states. For many realist scholars, there is little, if any, reason to assume that states' behavior would change even though the arena of this interaction is now in outer space, which, as will be demonstrated in Chapter 2, is precisely the way it occurred during the formative years of the modern space age.

Thus, for many realists, space should be regarded as merely the newest arena in the long-standing, traditional competition among states vying for the opportunities to assure their national security in an anarchic system. Space adds another dimension to modern competitive world politics, which has been likened to a three-dimensional chessboard consisting of unipolar, bipolar, and multipolar facets.[16] The idea of space being an extension of terrestrial geopolitics has received the most scholarly attention, though finding an acceptable parallel to properly develop a theory of space power has proven contentious and elusive.[17] Nonetheless, the role of space programs as an integral element of state power follows in a long tradition of the primacy of national security and power, areas in which international relations theory has evolved to explain states' actions in terrestrial matters.

Power, as understood with the context of international relations theory, is the manifest ability of an actor to exert influence over other actors in the international system, which as Hans Morgenthau (1948) argued, was the natural trajectory of states.[18] A theory of space power, therefore, builds upon the traditional concepts of international relations theory, but applies the history and attendant realities of space to understanding the policies and motives of modern space-faring states. Foremost in the realist understanding of state actions has been the theoretical model of geopolitics. While theorists differ on some of the specifics, geopolitics is typically concerned with the relationships between states, proposing a "problem-solving theory for the conceptualization and practice of statecraft."[19] First proposed by the Swedish political scientist Johan Rudolf Kjellén in 1899, geopolitics is among the oldest recognized political approaches, emphasizing the geographical relationship between states and the contexts through which their power is applied and utilized.

The work of British geographer Halford Mackinder is frequently noted by space power theorists as being pivotal in modern geopolitics, which Mackinder called "the relationship of international political power to the geographical setting."[20] Geopolitics became a tool and policy justification for the European states' scramble for African colonies and for Nazi Germany's expansionist policies in the 1930s. Though generally discredited following World War II, geopolitics reemerged as a legitimate tool for policy analysis during the 1970s in United States and Europe, and remained a cornerstone for strategizing among major states during the Cold War.[21] It fits logically that some parallels can be drawn from the lessons of terrestrial geopolitics that would inform us how we might expect the politics of space to evolve, given the fact that states are still pursuing power, security, and development in an anarchical international system, even though the arena of state interaction now extends into space.

The international relations theorist Hedley Bull observed during the early stages of the space race that "the first missile powers contemplate space with the perspective of the first oceanic naval powers ... and their experience provides them only with analogies."[22] More recently, Everett Dolman (2001) has extended this logic to argue that the state that controls the orbital pathways will be able to dominate terrestrial matters as well. Dolman argues that space power is an extension of the theories of realism and geopolitics that have evolved over the past 150 years, and that the evolution of a theory of space power follows directly in the footsteps of all previous theories about the development of national power, having no extraordinary differences.[23]

Other areas also offer examples to inform an emerging theory of space power. Charles de Gaulle once opined that war was the creator and destroyer of states. Technological innovation by states has almost unfailingly been driven by the need to ensure their security against foreign aggression. Political aims coupled with the quest for industrial power and sophistication have for over a millennium formed what William H. McNeill (1982) termed the "pursuit of power."[24] Railroads, for example, were the first major technological innovation that directly impacted states' capacity to exert command and control over a broad geographical area, to exploit far-flung natural resources, and to build vibrant trade-based economies. Burgeoning late-nineteenth-century powers such as Argentina, Britain, United States, and Russia all built railroads, not only to span their countries but also to control them militarily and exploit the countries' economic potential and natural resources. But, as Norman Friedman notes, railroads were of limited geostrategic utility given their immobility.[25] Extending from this failing, the expanse of the oceans became the next great geostrategic arena in which states would try to exercise control.

Scholars generally acknowledge Alfred Thayer Mahan's *The Influence of Sea Power upon History* (1890) as the first great modern treatise to expound upon the importance of great powers controlling the world's oceans as a foundation of their economic and military strength.[26] Mahan argued that the world's oceans comprised "a great highway," and the ability of great powers to control this thoroughfare had always been the foundation of their security. This work was fundamental in shaping strategic thinking in the United States and other countries about the importance of naval power for defense and trade, the protection of which is collectively known today as "national security." Based in no small part on this powerful message, the United States first constructed its vast navy, elements of which (16 battleships) Teddy Roosevelt paraded around the world in 1909 as the "Great White Fleet" to demonstrate American military power and to inspire awe, command respect, and build prestige vis-à-vis both potential allies and adversaries. Other countries swiftly followed suit in constructing large "blue water" navies.

Just as states previously recognized the geostrategic necessity and utility of railroads and navies, today's current and aspiring space-capable countries have added space technology as a vital element in the determination of economic and political security in the modern state system. Space programs fit well into the understanding of state power because they simultaneously provide aspects of both hard and soft power. By their nature, space programs are an archetypal example of the development of hard power, since they require the creation and utilization of many multi-use technologies that expand a state's capabilities in security. For example, the same propulsion technology that can put a satellite into orbit can likewise propel an explosive payload at an enemy. This type of technology also typically spawns and improves other associated technologies, as well as improving the technical skill of the state's scientific base. In addition, space programs provide multiple venues for soft power, such as building national pride and international prestige as well as creating an image to other actors of increased capabilities and potential, perhaps to be emulated or followed. Such was the case for the first states to develop space programs, which on the one hand yielded many tangible benefits due to interstate competition, and on the other hand laid the foundation for a considerable amount of the struggle of the Cold War. The development of space technology offered the superpowers powerful and nuanced strategic options.

Space programs bestow equally important soft power, especially those that involve human space flight. Every major space power has spent considerable funds to achieve the ability to put humans in space for both tangible and intangible benefits. Logsdon (2007) has argued that human space flight ranks among the most intensely patriotic symbols of modern times.[27] Some of the emerging space actors have pursued or are pursuing human space flight as a demonstration of their programs' sophistication, and their astronauts are held up by their governments as national patriotic icons. As will be discussed in Chapter 3, for the largest EMSAs—Brazil, China, and India—their space programs have been touted not only as national accomplishments but as a national catharsis to overcome histories of direct and indirect domination by outside powers and to project to others a sense of greatness.

For almost two decades, the space race occupied a central place in the national security strategies of the major powers, which realists argue is only natural and expected because states were merely doing what states do—trying to ensure their survival. Dolman (2002) has incisively observed the classic balance-of-power argument in space policy: "it was [the] pattern of perceived military necessity shouldered for fear of the growing power of a potential enemy that ultimately drove the development of space programs."[28] Deudney (1991) not only presaged this argument, but also noted that the emphasis on state competition in space has been beneficial, since it was the *realpolitik* of the Cold War that brought about technological advances in space technology far ahead of what would have been expected absent interstate conflict.[29]

It logically follows, therefore, that either directly or indirectly the lion's share of state investment in the larger, more established space programs in wealthier countries has followed historical precedents by emphasizing national security, traditionally focusing on military matters. Beginning in the late 1950s, military projects accounted for between 75 and 90 percent of all US spending on space.[30] While it might be argued that the Apollo program changed that spending priority, the moon program itself was very much driven by international competition and the aforementioned quest for prestige.

Current funding levels for military space programs in the United States are much obscured, since space-related activities are typically not listed as separate line items in the Department of Defense (DoD) budget. The best estimates put the total (DoD FY2010–2011 space budget classified and unclassified) at US$80 billion.[31] In Russia, after a long budgetary winter following the Soviet Union's fall, the Russian state space program budget rebounded in 2006 to 305 billion rubles (US$11.7 billion).[32] As will be illustrated in Chapter 3, the birth and evolution of the space programs in the larger developing states—Brazil, China, and India—has, to date, followed a parallel trajectory to that of their more developed counterparts; national defense concerns (albeit emerging later in India), directly or indirectly, have been the most important, if not exclusive, impetus behind their space programs, notwithstanding rhetoric to the contrary. Even for the less well endowed EMSAs, space programs frequently play a key role in national security, though typically focusing as much, if not more, on socio-economic development.

The commons of space

The counterbalance to the assumption of competition in states' space activities is derived from international liberalism, which postulates that international cooperation can occur provided the existence of the right institution, regimes, and norms, even if, as is usually the case, states' interests differ. The United States, as one of the two founding states of the space age, set a laudable, public precedent in international cooperation in the founding of the National Aeronautics and Space Administration (NASA). The language contained in its founding charter defined the organization's goal as promoting "peaceful activities for the benefit of mankind" and engaging "in a program of international cooperation" as directed by the US president (though, as will be discussed in Chapter 2, security concerns quietly played an equal founding role).[33] The imperative for cooperation in space occurs, generally, because it benefits the state, and specifically, because of the very high demands on a state's financial, scientific, and technological capabilities to produce a space program.

International cooperation can expand the range of an individual state's program by utilizing the scientific, technological, and resource capacities of other states. While the formative years of the space age were less cooperative, interstate cooperation has become much more common in this second era of space programs, though not always necessarily for the good of the commons.

While realism declares that conflict is the norm in international relations, liberalism counters that there is cooperation despite anarchy, and assumes that states' interactions in economic, cultural, and intergovernmental fora allow *all* states to achieve relative gains (positive-sum), whereas realism envisions only a winner-take-all situation (zero-sum). These examples of cooperation are called regimes, and the most commonly cited definition of regimes comes from Stephen Krasner (1983), who defines them as "institutions possessing norms, decisions, rules, and procedures which facilitate a *convergence of expectations* [author's emphasis] in a given area."[34] Given the absence of a Hobbesian Leviathan to bring order to the international system, regimes are, by definition, examples of international cooperation. Wherever there are areas in which the interests of two or more states overlap, in the absence of conflict, cooperative regimes may emerge, intended to achieve and ensure cooperation and shared benefits.

States have increasingly established regimes to govern interstate activity in an international system where the defining characteristic is still state sovereignty. Within the liberal tradition of international relations theory, regimes emerge to coordinate actions in the areas in which the interests of two or more states overlap. Since World War II, the growth of cooperative regimes has been exceptional, and these regimes have manifested themselves in a multitude of forms and purposes, including trade agreements (e.g., the GATT), international cooperation (e.g., INTERPOL), intergovernmental institutions (e.g., the United Nations), human rights (e.g., the Geneva Conventions), and collective security (e.g., NATO). These regimes serve to promote international cooperation and to regulate state behavior in policy areas outside of the realm of state sovereignty through mutually profitable arrangements and compromises by states for the common good. By extension, the establishment of regimes to govern states' interaction in space follows a natural progression founded on previous cooperative arrangements.

In the pursuit of peace and cooperation in space, regimes have been held up as the key to addressing the lack of concrete demarcation and agreement on sovereignty over such areas as orbital paths and celestial bodies. Sovereignty has traditionally been well understood and accepted where fixed borders have existed. However, the murkiness of the limits of sovereignty has been confronted before in international treaties governing the open oceans and airspace, which both remain areas where states' interests and the anarchy of the international system meet, flounder about each other, and have eventually found some limited common ground to establish a framework for international cooperation. Three brief examples illustrate how international regimes have turned previous areas of international anarchy into ones of international cooperation, or have at least prevented conflict. The roots of international cooperation in space are found in maritime law, air law, and international agreements on shared commons, such as Antarctica.

The world's oceans, making up 70 percent of the Earth's surface, had for most of recorded history been a political no-man's-land. Only in 1958 did the United Nations Convention on the Law of the Seas (UNCLOS) create a basis for demarcating states' sovereignty claims in the ocean. The criterion for creating "territoriality" in the ocean was a state's universally recognized coast as the base point for determination of sea sovereignty. However, beyond the 200-nautical-mile Exclusive Economic Zone (which only 25 states so far claim), the world's oceans remain an outpost of anarchy, with international shipping regulations that are normally façades, rendering the world's oceans subject to the whims and interests of states and non-state actors alike.[35]

A similar illustration of the extent and limits of state sovereignty may be seen in the development of international air law after the invention of the airplane. The French jurist Paul Fauchille forwarded the proposal that was to determine the extent to which civilian aircraft from one country could fly over other states.[36] Although at the beginning of the aircraft age sovereignty was understood to extend to the atmosphere above a given state (*usque ad coelum*), subsequent international agreements created the right of innocent passage, which is still coordinated by the UN International Civil Aviation Organization.

Lastly, the situation of Antarctica offers important parallels to outer space and possible approaches to its cooperative use via international agreements. As the last unclaimed land mass on Earth, Antarctica presented a particularly special dilemma and mirrors some of the incongruities and proposed solutions for the unique problems of space vis-à-vis terrestrial states' interests. As Antarctica was a potential source of international tension during the Cold War, negotiations between 12 states with interests in the frozen continent produced the Antarctic Treaty of 1959. This agreement proscribed territorial claims, prohibited the use of Antarctica for military activity, and affirmed the cooperative, peaceful use of the continent for scientific research. Though seven states have so far made territorial claims to Antarctica, none is internationally recognized.

Each of these examples has been proposed as a template for cooperation in space, laying the groundwork for the creation and evolution of space law, which attempts to extend these existent precedents of international law governing state activities into space. Some aspects of space law are mundane and perfunctionary, such as determining where the atmosphere ends and space begins (generally agreed to be the so-called Kármán Line at an altitude of 100 kilometers).[37] By contrast, most of space law has wrestled with the same substantive issues of nuclear weapons, conflict, and resource claims and ownership that emerged in the other previously discussed "commons areas." This process was formalized in the creation of the United Nations Committee on the Peaceful Uses of Space (COPUOS) in 1959, which established a forum for international debate and the negotiation of rules to govern space activities.

The concept of space as a commons of humanity is a popular one, typical in high-profile space missions (e.g., the International Space Station) that portray space as a place where humanity works cooperatively to promote exploration and "pure science." This attitude is reflected in the United States, where public support for the most visible, if not benign, aspect of the US space program—the shuttle program—has consistently remained high with 82 percent of the US population supporting it, even in the wake of the 2003 *Columbia* disaster.[38] This level of support has remained largely unchanged from the early days of space race.[39] While marginally lower, European support for similar programs has remained steady as well. This communitarian view of space has been affirmed at the level of international law, which has attempted to codify and standardize conduct in and use of outer space for the common good and the prevention of conflict.

The evolution of space law has so far paralleled the space age. Within two years of the launch of *Sputnik* in 1957, the United Nations founded COPUOS, which generated five international treaties governing international cooperation in space and space activities. The two most important are the 1967 Treaty on Principles Governing the Activities of States in the Exploration and Use of Outer Space, Including the Moon and Other Celestial Bodies (the "Outer Space Treaty" or OST) and the 1979 Agreement Governing the Activities of States on the Moon and Other Celestial Bodies (the "Moon Treaty").

Based upon the legal concept of *res communis*, the OST defined the limits of permissible space activities and committed states to cooperative policies of space exploration "for the benefit and in the interests of all countries." The OST proscribes the testing or basing of weapons of mass destruction in space and affirms that "outer space, including the moon and other celestial bodies, is not subject to national appropriation by claim of sovereignty, by means of use or occupation, or by any other means." The OST has been a successful regime because, so far, the 125 states (as of 2008; 98 signatories and 27 signed but not ratified) have adhered to the treaty's stipulations.

Building upon the OST, the 1979 Moon Treaty was intended to establish a regime for the use of the moon and other celestial bodies. This treaty was explicitly fashioned to follow the precedents set forth in the Convention on the Law of the Sea, stipulating that jurisdiction of the moon and the use of its resources be undertaken only by the international community, thereby proscribing states' claims, in whole or in part. Unlike the OST, however, the moon Treaty is largely viewed as a failed regime. Though 13 countries have ratified it, none of them include current or aspiring space-faring powers, presumably because of the unknown parameters of the moon's natural resources, such as helium-3.[40]

Recognizing the future imperative of staking a claim to space for their national future, some developing countries have tried to create regimes to place their claim in space vis-à-vis more established space actors. For example, in 1976 eight equatorial countries—Ecuador, Colombia, Brazil, Congo, Zaire, Uganda, Kenya, and Indonesia—attempted to assert that the portion of geosynchronous orbit over their national territories belonged to them.[41] This so-called Bogotá Declaration was in direct contravention to the OST. This attempt, like other claims to geosynchronous orbits, has gone unrecognized by all other states, thus paralleling earlier Antarctic territorial claims.

However, while cooperation can and does occur, it has not been, and some realists argue will never be, the norm.[42] The assumption that space programs have been the products of cooperation among states is not reflected in the history of the development of state-sponsored space programs and is at odds with the long-held concept of state sovereignty as established by the 1648 Treaty of Westphalia. Thus far, all space-faring states have had a national security interest in utilizing space, and in the near term, the cooperative, conflict-free use of space seems unlikely because national space policies, particularly of the larger, more capable states, have been almost exclusively fashioned according to the tenets of realist competition.

This is not to say that there have not indeed been selected projects that have revealed humanity's cooperative potential (for example, the Hubble Space Telescope). But the historical record demonstrates that even when states undertake space projects that are presumably cooperative in nature, the true intention is normally to further the political, strategic, and economic goals of the individual state, and not necessarily to promote the "good of all mankind."[43]

That historical record, and the template it has created for up-and-coming space actors, was largely the product of the dawn of the space age and the East–West competition of the Cold War. It is in this competition that we will look for the seeds that have spawned today's burgeoning second space race among developing countries.

2 The evolution of national space policies

Per ardua, ad astra.
Virgil

As is inevitably the case, in order to understand a contemporary phenomenon, one must appreciate the contribution of its formative history. This chapter concisely examines the historical, political, and economic evolution of national space policies at the beginning of the space age, as a basis for understanding, categorizing, and comparing the space policies of developing countries. For emerging space actors (EMSAs), their path into space and space policy has been largely paved by the space-faring states that came before them, which established the practices, norms, and legal environment of space activities today. This chapter examines the genesis of space policy in the modern state system and analyzes how space programs evolved and assumed a place of policy prominence during the Cold War—first among the competing superpowers, and then among other significant developed states.

This chapter also provides a reflection on the concurrent development of missile and nuclear programs by these same powers, which is appropriate and necessary because of the dual-use nature of these technologies. A technological and political maxim that materialized during the space age is that there has been an inexorable and symbiotic relationship between space programs, missile technology, and nuclear programs, whenever technologically and politically feasible. This interlocking triad goes a long way in explaining the impetus for the creation of national space programs among the larger and wealthier countries of the developing world. The explanation for this symbiosis is straightforward: the same technology that can put satellites in space can also launch weapon payloads at an enemy, and in the post-World War II period, the ultimate expression of national security for larger states has been the development, or at least the threat of development, of rocketry and nuclear weapons.

Thus, a recurrent theme herein is the extent to which national security considerations, in terms of strategic and tactical gains as well as propaganda value, shaped and impelled the various stages of these first space programs. Socioeconomic benefits were frequently secondary considerations at best. These security benefits have also been argued as the rationale for the development of nuclear weapons.[1] The present study suggests that this motivation also largely prevails among the larger and more developed EMSAs today. While there have been many important achievements in space research for the purpose of "pure science" (e.g., the Hubble Space Telescope, the three Mars rovers, the Cassini probe, etc.) and a number of noteworthy international agreements to promote cooperation in space, the historical evidence patently demonstrates that the sustained peaceful and cooperative exploration of space was never considered a viable or even desirable option by the largest and most capable states during the space race, an outlook that persists on some level to this day among the larger emerging economic powers.[2] When scientific space-related research has been carried out, the impetus for it has tended to be either to provide the foundation for future military space endeavors or to match the scientific and propaganda efforts of others, thereby buttressing the realist argument that a balance-of-power mentality has imbued national space policies. The scope and the budgets of non-security-related space research have traditionally paled in comparison to the national-security-oriented efforts of the larger and more capable emerging space actors.

In short, much of the history of space exploration during the formative years of national space programs has been essentially that of the attempted control and, occasionally, the militarization of space. This established paradigm continues to influence the development of many, though not all, emerging space programs of the developing world, though these have the added motivation of economic and social development through improved communications and remote-sensing technologies.

Precursors to the space age

The beginning of the space age is akin to the onset of maturity. While there are obvious outward signs, there are also innumerable, more subtle, incremental precursors that lead to maturation. Such is the case with the space age. Though there were hallmark moments in the development of space flight, there were many smaller steps that led to space being considered as a part of national security policy.

The history of rocketry is longer than many imagine. The first self-propelled projectile, a steam-powered rocket, was designed and built in ancient Greece by Archytas of Tarentum around 400 BCE. The first known ballistic missiles appeared in the eleventh century CE when Chinese armies began using gunpowder to launch arrows in battle.[3] A number of early military experts and fiction writers in Europe speculatively described rocketry, perhaps most famously in Cyrano de Bergerac's *Histoire Comique des*

États et Empires de la Lune et du Soleil, a work in which the description of rocket flight is rooted in part in physics. But it was not until the publication of Isaac Newton's *Philosophiae Naturalis Principia Mathematica* (1687) that the mathematical models necessary to make modern rockets work were sufficiently described. Based on Newton's theoretical work, beginning in the eighteenth century engineers from a diverse collection of countries began experimenting with rocketry for their military arsenals.

In 1792, the Indian armies of Tipu Sultan launched rocket barrages against invading British forces during the Mysore Wars, which understandably piqued British interest in rocketry. Based on the pioneering experiments of the Englishman William Congreve, whose rockets had ranges of up to 1,000 meters, the British implemented the military use of "Congreve rockets": first in 1806 during the Napoleonic Wars, and then against the United States in 1814 (inspiring the "rockets' red glare" of the US national anthem). But in each case, the rockets' relatively small size (typically 10-kilogram warheads loaded with case-shot carbine balls) meant that their tactical impact was very limited. Instead, their value lay in their use as psychological "terror weapons" to confuse and demoralize the opposition.

During the nineteenth century, most European powers developed rocket manufacturing plants and established dedicated rocket brigades. Russia, for example, founded its first rocket plant in Saint Petersburg in 1826, which supplied Russian troops with rockets during the Russo-Turkish War of 1828–29. The United States later used British-produced Hale rockets, an improved spinning variant of the Congreve, against Mexican forces during the Mexican–American War (1846–48). But it would not be until the twentieth century that rocketry's potential contribution to national strategic policies would be fully appreciated and realized.

The fathers of modern rocketry

Working independently of each other but largely cognizant of each others' work, three men in the early twentieth century set the stage for modern rocketry and space flight. The modern pioneer of the astronautic theory was Russia's Konstantin Tsiolkovsky, who inspired future generations of rocket visionaries with his blueprints of the first modern rocket and forged a number of ideas crucial to space travel. Tsiolkovsky's rockets were designed to be multi-staged (necessary to reach orbit) and powered by liquid oxygen and liquid hydrogen, which would give them sufficient thrust to achieve the necessary escape velocity of 11.2 kilometers per second. Though he never saw his rockets achieve flight, his writings and designs are considered the essential foundation for the development of modern rocketry and space flight. Following Tsiolkovsky's death in 1935, Stalin's purges (1936–38) decimated the Soviet scientific community along with many other sectors of Soviet society, assuring that the next stage of rocket development would happen elsewhere.[4]

The first working liquid-fuel rocket engine was built by Peruvian scientist Pedro Paulet in 1895, though it was never attached to a rocket for lack of funds.[5] That achievement belonged to Robert Goddard, an American who designed and launched the world's first liquid-fuel rocket in 1926. Bolstered by a breakthrough steam-turbine nozzle invented by Sweden's Gustaf de Laval, Goddard's rockets achieved supersonic flight by the mid-1930s. But the application of Goddard's invention was thwarted because his proposals to the US Army for funding were rebuffed and because of his penchant for secrecy, in the belief that all liquid-fuel rockets were his proprietary invention.[6] With the exception of the US Army's aforementioned brief foray into rocketry during the Mexican–American War, the exploitation of the tactical and strategic potential of rockets had not yet been recognized by US policymakers as a "professional" venture; this attitude fits into the general early twentieth century pattern in the United States of ignoring all things aeronautical.

For example, at the outbreak of World War I in 1914, the US had only 23 obsolete aircraft, compared to Germany's then-modern 1,000 fighters. Neither the US Congress nor the US military were easily convinced that the benefits would be worth the cost.[7] This same deficit of prescience would reemerge for a time during the early space age. Nonetheless, through the sponsorship of the Smithsonian Institution and the Guggenheim family, Goddard was able to continue his work privately, conducting rocket experiments in the New Mexico desert. While he languished in obscurity in the United States, the importance of his work was recognized across the Atlantic Ocean.

The birth of the space age

The tangible beginning of the modern space age and the concomitant outgrowth of national space policy stem from the work of Germany's Wernher von Braun, who was a protégé of the German rocket designer Hermann Oberth. In producing his visionary rockets, von Braun openly incorporated Robert Goddard's research (which had been published by the Smithsonian Institution in 1919) into his own designs.[8] Von Braun and other German designers' experiments in rocketry were facilitated by a loophole in the Treaty of Versailles, the onerous peace accord imposed on Germany after World War I. Though the treaty had set severe restrictions on German military research and the number and type of armaments allowed to the German military, the victors of World War I had not anticipated the advent of rocketry and therefore had not addressed it in the treaty's language.

As an active member of the Verein für Raumschiffahrt (Society for Space Travel), von Braun led a group of rocket enthusiasts, who operated a test area near Berlin called the Raketenflugplatz (Rocket Airport).[9] Von Braun was stunned by the US government's inattention to rocketry and the lack of funding for Goddard's work.[10] But, ironically, it was von Braun's own lack of financing, together with the fact that civilian rocket tests were made illegal by the Nazis in the 1930s, that changed the course of his research. After being shut down, he made a Faustian bargain by joining the Nazi Party and later becoming an honorary major in the Schutzstaffel (SS), which allowed him to receive financial support for his rocket experiments. Accordingly, his doctoral dissertation, *Konstruktive, theoretische und experiementelle Beiträge zu dem Problem der Flüssigkeitsrakete* (Design, Theoretical, and Experimental Contributions to the Problem of the Liquid-Fuel Rocket), was classified "top secret" by the Nazis. Though he had dreamed of becoming the "Columbus of space," von Braun instead settled for becoming the director of the new Nazi rocket research facility near the village of Peenemünde on Germany's northern Baltic Sea coast.[11]

There, in 1932, the 24-year-old von Braun developed a rocket research program for the German army and produced an ever-improving series of rockets called Aggregate. Once World War II began, the rocket type was renamed as the now-infamous V-2 (Vergeltungswaffe or "Vengeance Weapon"). After the first successful test in October 1942, von Braun convinced Hitler to authorize mass production of the V-2 the following year. Eventually, over 6,000 V-2s were constructed and some 3,200 were launched against southern England as well as Belgium and France from September 1944 to March 1945. Despite von Braun's romantic notions of space travel, the German military had instead turned his invention into the world's first guided ballistic missile of war.

The most immediate effect of the V-2 was the fleetness of its impact. Winston Churchill once darkly described the V-2 as "the Angel of Death ... only you can't always hear the flutter of its wings."[12] The immediate predecessor of the V-2, the cruise missile prototype V-1 (colloquially known in Britain as the "buzz bomb" for its distinctive drone), had flown at a relatively sluggish 640 kph, which, in the end, was no match for the interlocking system of British radar and RAF Spitfires, which could reach speeds of over 800 kph in a dive. On the other hand, there was no defense against the V-2. Carrying one ton of high explosives, the V-2 ascended into the mesosphere to an altitude of over 80 kilometers before diving onto its target at approximately Mach 5 (5,760 kph). There was no warning of its impending impact. Von Braun later ruefully observed that the design worked perfectly, "except for landing on the wrong planet."[13] He was, in fact, briefly incarcerated by the Gestapo in 1944 for his assertion that the project was really meant to achieve space flight.[14]

But, in the end, the introduction of these breakthrough rockets was a case of too little, too late. Despite the use of over 3,000 V-1 and V-2 rockets which destroyed or damaged over 33,000 houses and killed over 7,000 people (2,754 in Britain alone), the rockets were not strategically significant and could not turn the tide of the war.[15] The rocket's impact on the victorious Allied leaders was, however, indelible. Von Braun's invention had demonstrated in no uncertain terms the military potential of ballistic missiles, especially when the Allies learned after the war of Germany's Amerikarakete, a planned intercontinental ballistic missile, and Projekt Amerika, a submarine-launched ballistic missile that was to have been used against the eastern United States, but that died on the drafting table.[16]

With victory within sight, American and Soviet policymakers turned their attention to planning for their respective postwar national security needs, and began to divert policies and resources from simply winning the war to pre-positioning themselves for the radically altered international system of the postwar period. Though they were still officially allies, unified temporarily to defeat Nazi Germany, the irreconcilable demarcation between the communist and capitalist worlds was becoming apparent. In this geopolitical shift, the acquisition of Germany's missile and nuclear technology was given the highest priority by both sides. Having seen the V-2 in action, and despite some holdouts wedded to the strategies of traditional airpower, a growing number of American and Soviet strategists began to realize the importance that ballistic missiles would play in the future balance of power. Whichever side could capture the most personnel and resources would have an advantage in the looming but yet to be named Cold War. The V-2, together with the recently manufactured atomic bomb, revolutionized not only rocketry but the entire paradigm of state security, laying the groundwork for all capable states' subsequent policies in defense, development, and technology. A new opportunity for achieving what both superpowers thought would be an unstoppable military deterrent was now available.

Accordingly, the previous lack of interest in rocketry on the part of the US military was replaced by a sense of urgency to develop the potential of rockets. The first halting attempt in the United States to create indigenous rocket programs began just before the end of the war. In November 1944, the US Army contracted General Electric to construct a variety of missiles, from short-range tactical ballistic missiles to intercontinental ballistic missiles (ICBMs) to surface-to-air (SAM) missiles. The US Navy also dabbled in rocketry with Operation Bumblebee, an attempt to create the first antiaircraft missile (similar to Germany's previous Wasserfall program). But the true impetus behind the policy shift was Operation Paperclip (originally designated Operation Overcast), the code name for the secret US program to find and enlist the services of erstwhile enemies by bringing captured German scientists and their rockets to the United States.[17] The British ran a similar program called 30AU, headed by Ian Fleming of future James Bond fame.

Von Braun's name was at the top of the United States' "Black List," which contained the names of prominent German and other Axis scientists and engineers whom the United States sought to acquire ahead of the Soviets' advance from the East. The explicit purpose was twofold:

acquiring technology and knowledge to jump-start the US rocket program while denying the Soviets access to the same. Concurrent with Operation Paperclip, both the United States and the Soviet Union ran parallel search programs to find and remove both personnel and whatever surviving records and equipment could be discovered on Germany's nuclear weapons research.[18] Located deep under a castle in the southwestern town of Haigerloch, German physicists had built a nuclear reactor to try to produce a nuclear chain reaction, in a project called Uranverein (Uranium Club). The US knew of this program and sent operatives of Operation Alsos to recover the equipment as well as the thousands of kilograms of uranium known to be in German hands. In addition to sharing these objectives, the Soviet Union's search program had the added feature of hoping to move entire production facilities back to the USSR. Soviet efforts, however, were hampered by a chaotic scramble for postwar spoils by competing Soviet industries and military departments.[19]

Seeing the certain end of the war and wishing to avoid capture by the Soviets, von Braun led a group of some 500 scientists, along with their Peenemünde records, to southern Germany where they surrendered to the US Army on 2 May 1945. Von Braun and 127 scientists were spirited out of Germany along with approximately 100 V-2s. Within a few months, von Braun and about 120 other German scientists were working for the US Army Ordnance Corps at the White Sands Proving Grounds in southern New Mexico.[20] Throughout the almost three decades of Operation Paperclip (1945–1973) over 1,600 European scientists would be relocated to the United States and Britain to help their respective rocket and nuclear programs.[21] The United States and other Western powers profited greatly from their human scientific plunder.

But despite the best efforts of the US to forestall the Soviet effort, the Red Army was successful in capturing most of the V-2 production facilities and rounding up its own cadre of some 2,000 German rocket engineers and technicians, including Helmut Gröttrup, Erich Putze, and Werner Baum, who were experts in guidance, production, and propulsion, respectively.[22] Along with the remains of the V-2 factory at Nordhausen that had not already been stolen by US forces, these scientists were taken behind the Iron Curtain, where they were employed toward realizing the same goals as their compatriots working in Britain and the United States. Under the direction of the lead Soviet rocket engineer, Sergei Korolev, the captured German engineers then reestablished the V-2 production facility in communist East Germany.[23]

In spite of the US Immigration and Nationality Act of 1952, which explicitly prohibited Nazi officials from immigrating to the US, von Braun and the other German scientists (around three-quarters of German scientists had belonged to the Nazi party) were all eventually offered US citizenship after their Nazi backgrounds had been sanitized by US intelligence agencies.[24] The larger issue of Cold War competition and the growing importance of ballistic missiles to national security completely overshadowed the engineers' former national allegiances or even possible war crimes. The potential for technological advancement offered by these scientists was too great a prize to relinquish, and their work was increasingly perceived by the US military and policymakers as an essential component of national defense. Such was the case, for example, of Arthur Rudolph, who had worked at the Nordhausen-Dora concentration camp.[25] He would later help to design the *Saturn V* booster that sent the *Apollo 11* crew to the moon.

The simultaneous intelligence operations of the US and the Soviet Union to acquire the tools of rocketry heralded the approaching bitter ideological struggle between the two emerging superpowers. The subsequent space race and the concurrent nuclear brinksmanship that resulted from the clash of these ideological antipodes became the primary instruments meant to demonstrate the superiority of each state. Thus, the first space policies emerged as an integrative element of the national security mania of the Cold War opponents, forging a permanent role for space in each state's national security policies.

In line with the primary focus of this book, it is essential to recognize that the desire to capitalize on advanced German aeronautical and nuclear technology and knowhow was not limited to the two Cold War superpowers. Aspiring regional powers such as Argentina, Brazil, India, and South Africa also offered refuge and resources to former German scientists to continue their work. Perhaps the most illustrative example occurred in Argentina during the first presidency of Juan Domingo Perón (1946–52). Perón welcomed German and Vichy French scientists and technicians to help design Argentina's ambitious nuclear and rocket programs and to build up the country's air force, projects which were meant to rectify a balance-of-power struggle with Argentina's perennial rival, Brazil (which had been the recipient of many then-modern armaments from the United States during the war). Kurt Tank, the German designer of the renowned Focke-Wulf 190 fighter plane, accepted an offer in 1947 to become the director of the Fábrica Militar de Aviones (Military Airplane Factory) in Córdoba, Argentina. There, under the alias of Pedro Matthies, Tank incorporated indigenous Argentine designs into his own and constructed the Pulqui II, one of the world's first operational jet fighters.[26] The following year, Tank's recommendation to Perón brought Austrian nuclear physicist Ronald Richter to Argentina to head the Huemul Project, a planned but failed nuclear fission reactor (though it is noteworthy as the first state-sponsored attempt at ostensibly peaceful nuclear energy production).[27] After Perón's overthrow in a 1955 military coup, Tank moved to India, where he designed that country's first jet fighter as well, the Hindustan Murat HF-24 fighter-bomber. There, Tank also taught aeronautical engineering to future Indian president and founder of India's ballistic missile and space programs, A.P.J. Abdul Kalam.[28]

In the United States, the V-2 rocket development program progressed apace as a private-public joint venture. The private sector, especially General Electric and Chrysler, helped to produce a new generation of American-made rockets.[29] The GE program used captured V-2s as templates, launching around 60 of them from the White Sands, New Mexico test site. The resulting American rocket that was proposed by GE in 1946 was the Hermes C1, which was to be the world's first multi-stage ballistic missile (it was downgraded to a single-stage and became the template for the later Redstone rocket).

After depleting the last of the captured V-2s, the von Braun team was moved in 1950 from New Mexico to the Army's new Redstone Arsenal missile center south of Huntsville, Alabama, the same year as the outbreak of the Korean War. With the US at war again, the missile development program took on renewed urgency. Von Braun received permission to develop the Redstone rocket, which had one objective: to carry an atomic warhead over 300 kilometers. However, it was the advent of the US hydrogen bomb (1,000 times more powerful than the Hiroshima weapon) in 1952 and the subsequent Soviet hydrogen bomb test in 1953 that spurred more substantive research funding into ballistic missiles. As Thucydides had put it some 2,400 years earlier in his *History of the Peloponnesian War* (1.88):

> The Spartans voted that the treaty [with Athens] had been broken, and war must be declared, not so much because they were persuaded by the arguments of their allies, as because they feared the growth of the power of the Athenians.

It was US policymakers' fear of the Soviet Union's growing capabilities in rocketry and nuclear weapons that finally opened up the coffers for accelerated research and development of ballistic missiles, which would, in turn, ultimately lay the foundation for the space program in the United States.

Emerging national space policies

The competition to design and build larger, more powerful rockets in the early Cold War period proceeded in parallel with the growth of nuclear weapons technology. Indelicately called a "balance of terror" by Winston Churchill, the resulting impasse came about because the two superpowers could not engage each other militarily without escalation to nuclear conflict. This zero-sum game forced the US and the Soviet Union into other avenues of competition, which normally found expression in less direct venues, such as proxy wars, alliance-building in the developing world, and, of course, competition in the space race.

In the first decade following World War II, Cold War tensions increased between the superpowers because of the expansion of communism worldwide. The successful communist takeovers in Eastern Europe and China, and the invasion of South Korea by the communist North in 1950, gave seemingly ample justification for the US anti-communist "domino theory," which purported that "losing" one country to communism made it more likely that others would follow. But notwithstanding the appearance of losing the ideological battle, the United States in fact maintained, and even increased, its decisive advantage over the Soviet Union in strategic bombers. Until Josef Stalin's death in 1953, that US advantage remained largely intact, with the United States actually gaining a significant tactical advantage by encircling the Soviet Union with nuclear-armed bombers (a situation that would not be seriously challenged until the 1962 Cuban Missile Crisis).

But the biggest obstacle that both superpowers faced during the early 1950s in putting into use their acquired rocket technology was that neither country had an established, institutionalized missile program that could adequately absorb and utilize the technology. Even so, the Soviet Union managed to establish a clear advantage in rocketry during the first postwar decade, aided in part by the streamlined policymaking apparatus in the USSR as well as by the fact that the United States had allowed itself to become complacent about new technology and innovation in the immediate postwar years, in part because of the superiority the US enjoyed in bombers.[30] This mismatch between policy and structure created an early barrier to innovation and advancement in missile systems within the United States. Unlike in the USSR, the US effort to establish missile and later space programs was full of detours, potholes, and frequent dead ends attributable to US political (i.e., electoral) realities, and the important influence of Air Force General Curtis LeMay, who prioritized strategic bombers over missiles. US missile policy was largely reactionary vis-à-vis the Soviet Union in the early years of the Cold War. The indecision of the US government about whether to invest in rocket and space programs stemmed in part from lingering differences among US policymakers regarding the importance that rockets should play in national security and how to best achieve those goals. This policy disconnect was largely absent in the Soviet Union, both among policymakers and among the public at large, who tended to view anything space-related with great enthusiasm.[31]

Equally problematic for the US was the fact that the design process of rocketry in the United States was highly disjointed due to the competitive nature of relations among the various military services in the first postwar decade. Until the creation of a civilian space organization in 1958, three of the four US military services—Army, Navy, and Air Force—competed against each other for funding and resources instead of pooling their efforts to build a unified rocket and space program.[32] During these crucial formative years, for example, the Army commissioned two dozen different missile projects. Not wanting the burgeoning field of rocketry to be dominated by the Army, the Navy started its own rocket program and trumped the Army by being the first to propose a satellite.[33] While the Army and Navy valued the development of their own independent rocket systems, their ultimate goal was not the conquest or even dominion of outer space, but to create a delivery system for the United States' growing nuclear deterrence capability, which myopically still focused on airplanes instead of long-range ballistic missiles.[34]

Complicating the situation in the US was the lack of political support for space-related programs, and such endeavors being subject to the political vagaries of the US democratic system, in which long-term goals are frequently subject to short-term electoral politics. For example, in 1947 President Truman vetoed an early proposal by Senator Alexander Smith for a national science organization (National Science Foundation Act S. 526) on the grounds that it would have given scientists sole spending discretion, therefore insulating the agency from "politics."[35] Also hampering missile development was the fact that Truman was faced with postwar inflationary fears, out-of-control federal spending, and the need to pay for the Marshall Plan as well as economic assistance to Japan. The budgetary squeeze forced Truman to make hard choices, but in light of the Air Force's predilection for proven airplanes over fanciful and unproven rockets, the Air Force's extant ballistic missile program, the MX-774, as well as ten other nascent missile projects proposed by the military branches, were cancelled. This left the United States without a serious and fully funded ballistic missile program until the Atlas program began in 1955.[36]

Despite these structural hurdles, the military services as well as private think tanks continued to press for the development of launch systems and satellites. The newly formed RAND Project (Research and Development; later renamed RAND Corporation) was most vociferous in its support and recognized early on the connection between missile and space programs. Under contract to the Douglas Aircraft Company, RAND issued a document in 1946 entitled *Preliminary Design of an Experimental World-Circling Spaceship*, which argued that "the development of a satellite [will] be directly applicable to the development of an intercontinental missile."[37] This document was the first authoritative, albeit sketchy, outline of a national space policy for the United States and in the world.

By comparison, the Soviet Union rather quickly embarked on a concerted rocket program. In 1946, a top-secret decree (#1017–419ss) issued by Stalin established the Soviet Union's rocket research industry by setting up a number of research institutes to take advantage of the recently acquired German technology. By the following year, the Soviet Union, with the assistance of its captured German engineers, was test-launching V-2s from a site near the Caspian Sea.[38] As a secret military program under the direction of the Ministry of Armaments, the Soviet rocket program expanded on earlier German concepts but had evolved a uniquely Russian character and design by the time the German scientists were abruptly repatriated back to Germany in the early to mid-1950s. By 1949, the Red Army had its first rocket division, supplied with the indigenous R-2 (an improved version of the captured V-2), which had a 600-kilometer range, double that of the V-2. So, from the beginning, the Soviet Union benefited from a fairly unified rocket policy that would facilitate an easier transition into its space program, though its implementation was not without its own impediments, such as initially dividing work unnecessarily between several competing design groups, as per Stalinist secrecy protocols.[39] Nonetheless, planning in the Soviet Union for the deployment of rockets was significantly ahead of that in the US during the first postwar decade.

So while the USSR moved decisively toward integrating rocketry into its strategic policy, the US during this crucial period had no formal rocket program, much less an official space policy or program. The only semblance of such organization was a civilian aviation agency, the National Advisory Committee for Aeronautics (NACA), whose primary concern was airplane design and performance. Even the US military entered into a brief period of confusion in rocket technology in the early postwar decade, rocket-research allocations having fallen precipitously.[40] Curiously, for a short time in the postwar period, United States rocket and space policy essentially resided in the tireless advocacy of Wernher von Braun, who was intent on "selling" outer space to policymakers in order to see his rockets fly.[41] Regardless of these circumstances, ballistic missiles soon proved themselves to be an indispensable contribution to national security and a symbol of twentieth-century power, progress, and prestige for the modern nation-state.

Turning plowshares into rockets

The first picture of Earth as seen from space was taken on 24 October 1946 from a V-2 test rocket launched from the Ordnance Proving Ground at White Sands, New Mexico. From an altitude of 100 kilometers, the aesthetically unpleasing, grainy black-and-white photos of the Earth's curvature demonstrated in no uncertain terms the tantalizing potential of the high ground of space for surveillance, mapping, and meteorology (a year later James Van Allen attached a Geiger-counter experiment to an updated Redstone that led to an understanding of the Earth's radiation belts). The urgency for developing the lift capabilities of these early rockets was heightened by the successful test of the Soviet's first nuclear weapon, a 22-kiloton bomb (code-named RDS-1), in August 1949. Thus, geostrategic political events played a crucial role at every turn in goading, persuading, or simply pushing the leadership of both countries into policy decisions that would further and deepen their commitment to and dependence on space as a tool of foreign policy. The 1950s saw rapid growth in the superpowers' rocket capabilities as well as important shifts in policy that made their later space programs integral to their respective national security goals in the Cold War. The heating up of the Cold War in the 1950s created a heightened sense of urgency in the superpowers to "conquer" space and, consequently, turned space into the center stage of the conflict.

US indecision surrounding its rocket programs started to wane as Cold War proxy conflicts became more menacing. The first Soviet nuclear weapon test in 1949 and the outbreak of the Korean War the following year prompted President Truman to change course. Three weeks after the North Korean invasion of South Korea, Truman commissioned von Braun's team to build the US ballistic missile Redstone.[42] Nonetheless, the US financial commitment to rocket research was at that period less than impressive. In the first postwar decade, US funding for ballistic missile research was anemic and, unbeknownst to Americans at the time, behind that of the Soviet Union, which had already designed a single-stage rocket called the R-14, capable of flying over 3,000 kilometers with a 3,000-kilogram nuclear warhead (but not ready for deployment until 1959).[43]

Part of the way in which the United States tried to play catch-up was to integrate work on its rocket programs, principally through the separation of military and civilian programs. The policy shift in the United States toward a military-civilian division of labor in rocket and space programs had a distinctly political purpose in that it reflected Eisenhower's desire to publicly distance the US military from space-related research efforts in early Cold War propaganda, so as to differentiate the US effort from the blatantly military space program of the Soviet Union. But creating separate programs caused a further division of resources. This distraction did not exist in the Soviet Union since its program was unabashedly military in the service of its foreign policy goals. Estimates by the CIA for the period 1945–57 put Soviet military spending at about 20 percent of its national product, compared to around 14 percent for the United States.[44] Though no specific figures are available for the formative years of the Soviet space program, the best estimates published by the CIA are that Soviet space programs were allocated at least US$3.4 billion (US$25 billion in 2010 dollars) annually by the 1960s.[45] Moreover, the Soviets were able to get a head start because their ambitious drive to build rocket and space programs was not matched in the United States, where space ambitions were running headlong into bureaucratic red tape and indecision among key policymakers.

The public perception of the goals of the American space program was another key reason for the early divide. It was a matter of concern and debate in the United States, but less so in the Soviet Union, which never pretended that its space program had a "softer" civilian side. The US policy direction was to create two faces for space activities: one that was overtly military and fit neatly into the established paradigm of national security, and one civilian, which was to undertake space projects for the ends of space exploration and "pure science." This differentiation has its roots in the development of the Vanguard rocket as the first US rocket specifically to launch satellites. This project included, for the first time, civilian management of booster and satellite development. During the period of 1946–52, the US invested less than US$1 million annually in rocket research in the Navaho guided missile program. By the end of the Korean War in 1953, investment had begun to rise slightly, though only to US$3 million, largely due to the Atlas program, which would produce the United States' first intercontinental ballistic missile.

Again, news from the Soviet Union prompted reactionary decision-making in Washington. In 1953, the German scientists who had been repatriated from the USSR back to Germany reported that the Soviets were developing a ballistic missile with a range of 3,700 kilometers, and that it would be operational by 1957.[46] This intelligence motivated the CIA to begin monitoring Soviet missile tests from covert sites in Turkey. The news was proven to be doubly worrisome after the Soviet Union tested its own hydrogen bomb in August 1953. For the first time since the war's end, the two superpowers seemed evenly matched in the strategic area of nuclear weaponry, and the Soviets were apparently pulling ahead in missile technology.

The looming specter of an evenly matched Soviet Union provoked a substantial increase in the United States in missile research funding. Beginning in 1955, the budget allocation rose to US$161 million and finally reached over US$1 billion in 1957, though this amount was still spread across six different missile programs, which themselves were divided among three different branches of the military.[47] But even with growing interest and research in finally completing a space-worthy launch vehicle, the US policymaking community continued to express pessimism about the ability of the United States to put a satellite, much less humans, in space.[48] While the civilian leadership was undecided as to the future of US space policy, the military branches were at last convinced of the merit of a space program. At the core of the matter, it was this disparity of vision that directly contributed to the United States lagging behind the USSR in the race toward space.

Dual-use policies

A key geopolitical moment in the space race that illustrates the contending space policy impulses behind even these earliest space programs was the establishment of the International Geophysical Year (IGY). Conceived in 1950 at the home of James Van Allen, the IGY was declared for 1957–58, which coincided with a period of maximum solar activity, and was meant to follow in the footsteps of the previous programs of the International Polar Years of 1882–83 and 1932–33.[49] The stated purpose of these scientific endeavors was to promote worldwide scientific cooperation through coordinated observations of an assortment of geophysical phenomena, such as the oceans, Antarctica, and geology. Scientists in the US nearly unanimously incorporated the IGY concept into their work, as did many of their Soviet counterparts.

What was novel about the IGY was the new and prominent role that space had assumed. The growing advances in rocketry by the superpowers had opened the upper atmosphere and the beginnings of outer space itself to more detailed exploration than previously attainable. Eventually 67 countries agreed to participate, the United States and Soviet Union among them, though their motives were more complex and strategic than implied by the program's benign public façade. Though the American scientists involved in the IGY were interested in scientific gains, the US government became involved because many of these scientists had government links through contracts, consulting, and related activities. The National Academy of Sciences recommended in 1954 that the US launch a satellite as part of the IGY, and in July 1955 Eisenhower announced this goal, which reflected the recognition by technocrats in the Eisenhower administration that satellites and space in general were an important new innovation with great potential for prestige-building, propaganda, and espionage.[50]

The two superpowers took very different paths to developing satellite programs for the IGY. The Soviet Union did not feign a civilian space program and unabashedly used military launchers for its contribution to the IGY. On the other hand, the US stressed the scientific image to establish the principle of overflight for civilian purposes. The Vanguard satellite program was the last in the pre-*Sputnik* efforts by the United States. At every stage of Vanguard's development, the overriding goal remained establishing the regime of overflight, though publicly the program's scientific and civilian nature was emphasized.[51] From the activity surrounding the IGY, the first national space policy of the United States emerged, albeit surreptitiously. The previous failure of the US to foresee the extent of Soviet nuclear weapons development, as well as the outbreak of the Korean War, illustrated the need for better intelligence of Soviet strategic capabilities and intentions. Issued in 1955, a then-secret National Security Council report, "US Scientific Satellite Program" (NSC 5520), contained recommendations that led directly to the production of the first Intermediate Range Ballistic Missile (IRBM), the construction of spy satellites, and the building of the U-2 spy plane—all instruments that would play crucial, pivotal roles in the Cold War.[52]

NSC 5520 also became the guiding policy for the practical national security benefits of the incipient space program. The document made two important points: first, it declared that the immediate purpose of US participation in the IGY was to test the limits of the "freedom of space" in order to establish a regime for future surveillance overflights of American spy satellites; second, it boldly touted the "prestige and psychological benefits [that would] accrue to the nation which first is successful in launching a satellite." The latter observation was a result of the CIA's contribution to the document and to the soon-to-born space program.[53]

Accordingly, even at this early stage, both the hard- and soft-power aspects of space applications were carefully weighed, and the IGY was considered to be the ideal setting for the launch of a satellite, providing the maximum propaganda benefit while using the IGY's scientific character to mask such an overt propaganda maneuver. NSC 5520 came to represent a recognized division, at least on paper, between civilian and military space programs: civilian programs were essential to establish precedents, after which military programs would quietly follow. The later creation of the National Aeronautics and Space Administration (NASA) was, in part, to provide a veil behind which military space programs could develop technology, techniques, and experience without the inconvenient public scrutiny associated with military ventures in democratic societies.

The overflight provision was recognized by the Eisenhower administration as being strategically crucial for the later deployment of intelligence-gathering satellites.[54] To further his objective, Eisenhower publicly announced his "Open Skies" initiative on 21 July 1955 at a summit conference held in Geneva, Switzerland, where he recommended that the US and Soviet Union allow overflights of each other's territory to reduce the fear and possibility of a surprise attack. But the true intention of Open Skies was to establish the legal precedent of the freedom of space, which was deemed imperative by US policymakers because if the Soviets were able to claim that national airspace extended into space, satellite surveillance would be mired in diplomatic wrangling for years. The Soviets rejected Eisenhower's proposal out-of-hand, though strategically, the president never expected the proposal to be taken seriously; it was, instead, a ploy to stump the Soviet Union by proposing a policy it would never accept, thus making it appear uncooperative, if not belligerent, to potential US allies in the developing world.[55] Only a decade after World War II, space politics had already become firmed entwined in the machinations of Cold War politics.

The United States could have easily been the first in space had it not been for the space program's image as perceived by the US public and legislators. In 1954 von Braun proposed Project Orbiter to the US government, requesting a paltry US$100,000 in funding to modify Redstone solid-fuel rockets in order to quickly put a satellite in orbit. His request was denied without comment, though it was probably because Redstone's military credentials might have cast US efforts in a bad light during the IGY, as well as because of the ongoing competition from rival US Navy and Army projects.[56] The Navy's Viking rocket, fitted with the new X-405 lox/kerosene engine as well as upper stages, finally got approval under the assumed name of Vanguard. Nevertheless, the US still was not yet ready to commit to the wholesale development of a space program, and allocated a mere US$3 million to satellite research.[57]

At the same time that Washington publicly touted peaceful cooperative science as the rationale for satellites, key civilian rocket and space scientists behind the IGY were an increasing presence at the president's National Security Council meetings.[58] The tactic employed was the classic chess maneuver of feigning with a pawn so as to, in reality, position the queen—in this case, portraying space as a venue for worldwide cooperation while really intending to politically and militarily outmaneuver the Soviets. This promotion of international liberalism to obfuscate patently realist policies would become a tactic regularly utilized by both sides during the space race.

The space competition heated up in 1955 after the United States officially announced its intention to launch a satellite as part of the IGY, prompting the Soviet Union to reciprocate. Nonetheless, the US effort remained woefully underfunded. However, there were some policymakers in the US who recognized the emerging connection of space power to both national security and international influence. Nelson Rockefeller, an advisor to President Eisenhower, encouraged the president to take the coming space race more seriously, complaining that Soviet success in their space program would have a dramatic effect on US prestige and could have negative consequences on "the political determination of free world countries to resist communist threats."[59] Eisenhower was initially unmoved; his eventual change of heart toward the use of space for national security can probably be attributed to the so-called Killian Report (produced in February 1955), a scientific study that outlined the potential uses of satellites as well as the Soviet ability to initiate a surprise attack on the United States with ballistic missiles and the imperative that the US develop a 1,500-mile intermediate range ballistic missile (IRBM).[60] The report also pointedly noted the political significance of the US being first in space, not only in terms of establishing the regime of overflight but also because of the "prestige and psychological benefits" of being the first country in space, especially in winning (and keeping) allies in the developing world.

Though Eisenhower acknowledged the psychological advantage and the attendant "prestige factor" a satellite launch would generate, he continued to resist releasing substantial funding for Vanguard because of repeated cost overruns. Consequently, most of the program ended up being paid for directly from the Department of Defense budget.[61] Even on the eve of *Sputnik*, Eisenhower only grudgingly increased funding and was never fully taken with the idea of prestige-building taking precedence over the tangible, tactical advantage of ballistic missiles and the surveillance satellites they would carry. The US Air Force, on the other hand, remained dubious as to the value of satellite photos in producing valuable intelligence, preferring to concentrate on ballistic missiles for nuclear payloads.[62] This quibbling ended, however, with the ignominious explosion of the new Vanguard rocket in December 1957, and with it, the possibility of the United States reaching space first.

By contrast, the Soviet Union had embraced rocketry from the very beginning, integrating it into the structure of its national policy. As early as 1946, a special military branch, the Supreme High Command Reserve (RVGK), was designated to first test captured V-2s and then assume the sole control over what later became known as the Strategic Rocket Forces (RVSN). Following Stalin's death in 1953, the new Soviet leader, Nikita Khrushchev, drastically shifted Soviet policy. An enthusiastic proponent of science, Khrushchev forcefully argued that missiles were the weapons of the future.[63] Khrushchev's faith in the value of ballistic missiles over conventional weapons was deepened by the 1956 Suez Crisis since it was only the threat of nuclear confrontation with the Soviet Union that forced the United States to pressure its allies, Britain and France, to withdraw.[64] Consequently, Soviet missile policy, and subsequent space policy, became predicated not only on their defensive utility but also their value as instruments of propaganda to deceive the West about Soviet intentions and abilities. In his own memoir, Khrushchev laid out this deception, saying that he specifically wanted to "give our enemy pause" by boldly claiming that the USSR could "shoot a fly out of space."[65]

This redoubling of the Soviet missile program meant an increased budget for Soviet launch facilities, including the construction of two testing complexes (a third facility in the Soviet Far East was planned but later cancelled). The larger and more important of the two facilities was founded in 1955 as a long-range missile complex on the southern desert steppes of the then-Soviet Republic of Kazakhstan. A few years later the complex was expanded to include launch facilities, and was given the name Baikonur Cosmodrome—the Soviet Union's spaceport.

The shot heard 'round the world

To the surprise of no one knowledgeable but of almost everyone else in the West, the Soviet Union fired the opening salvo of the space race. The launching of *Sputnik*, the world's first artificial satellite, on 4 October 1957 became the first in a long list of Soviet "space firsts" in the late 1950s that shifted the space race into overdrive and elevated the importance of space politics. Each space-related incident during this crucial period would further accentuate the growing rift between the ideological rivals, and would spur United States foreign and domestic policy into previously unexplored areas. These changes set the stage for a decade of bold, and occasionally outlandish, goals for the respective space programs. But, most importantly, the space-oriented competition consolidated space as a primary arena for international competition, national security, and economic advantage—a position of importance it has not since yielded.

Circling the Earth once every 96 minutes and emitting only a feeble repeating beep, the 70-kilogram *Sputnik* was a monumental political propaganda coup for the Soviet Union. The conventional wisdom in the

United States had been that the West was somehow innately superior in all areas, but especially in technology, and that the USSR was a backward society of food shortages, *babushkas*, and clunky machines.[66] Though this was largely a surprise in the West, the fact was that the USSR was indeed ahead of the United States in several key technological areas. The Soviet advantage was based on two major factors. First, the Soviet military settled on the development of ballistic missiles from the beginning, and chose not to pursue the early guided winged missiles initially favored by the Americans (e.g., the Navaho program).[67] As a result, by the time of *Sputnik*'s launch, the Soviets were at least a decade ahead of the US in ballistic missile systems. Second, many American strategists had undervalued space as an asset worth pursuing as a primary field of national security.

However, the response of the US government was not as reactionary as might be imagined, in part because the administration did not share the American public's reaction to *Sputnik*. While this reaction was vociferous, oscillating between shock and panic, the United States government's initial public reaction was surprisingly meek. This docility was attributable, to some extent, to the fact that President Eisenhower and others did not believe that *Sputnik* was a threat that required a rash response. Publicly, the official US reaction was underwhelming. General Curtis LeMay, head of US Strategic Air Command, called *Sputnik* "a hunk of iron" and dismissed its importance, calling it a "small ball in the air" and a "neat scientific trick," which was not to be worried about.[68]

On the other hand, Senate majority leader Lyndon B. Johnson used the occasion for political gain, criticizing Eisenhower's foreign policy by publicly opining that *Sputnik* signaled the beginning of Soviet domination of the "high ground" of space, and declaring that "control of space means control of the world."[69] Eisenhower, on the other hand, maintained a calm exterior, even issuing a statement congratulating the Soviet achievement. But underneath Eisenhower's congenial façade there existed an ulterior motive. Both the US intelligence agencies and the president himself had been struggling with the possible Soviet reactions to an overflight of its territory by an American satellite. But with *Sputnik*, the Soviet Union had solved that dilemma, and had established the precedent that outer space was not to be included in the time-tested territorial limits of national sovereignty. The Soviets had unintentionally helped to establish the concept of freedom of international space.

The other part of the explanation is that while Eisenhower held a deep trust in science, he had been slow to appreciate the broad-reaching implications of space, in part because of his perhaps ironic but deep misgivings about excessive military power. Eisenhower was determined to take a paced, reasoned approach to establishing a proper civilian organization for space activities. Though it has been argued that he still did not truly comprehend the momentousness of the Soviet accomplishment, Eisenhower nonetheless ordered the CIA to build spy planes and satellites so as to be prepared should the Soviets use this temporary advantage for a surprise attack on the United States.[70]

One of the most long-lasting responses to *Sputnik* came from the US Air Force, which quickly embarked upon a program called Man in Space Soonest (MISS), which was to have put an astronaut in space atop an Air Force Atlas rocket at the earliest possible date. Although MISS was cancelled by the Eisenhower administration, the Air Force continued its research.[71] The program would be replaced by NASA's Project Mercury, which would put the first US astronaut, John Glenn, into orbit. Though unable to immediately put a man in space in response to *Sputnik*, the decision was made to rush into orbit the first US satellite, *Explorer I*, on 31 January 1958.

Even faced with such formidable competition, Eisenhower would still not commit to the paradigm of a space race and its geopolitical implications: he tried to put a non-competitive spin on Explorer's launch by publicly reminding the US public that it was part of US participation in the IGY.[72] Eisenhower remained reluctant to fund space projects generously, probably due to his own military philosophy, which called for extremely careful planned and considered actions, as well as his concern that the United States must not be seen as following the Soviet Union's lead, but pursuing its own agenda.[73] For Eisenhower, the power of prestige in space would emanate from US leadership in space, and not from the image of a country desperately playing catch-up with the Soviet Union. But *Sputnik* did energize the US space program to attempt projects on a scale that would have otherwise not been attempted. Thus, *Sputnik* goaded the Eisenhower administration into finally acknowledging a necessity for an American presence in space and taking the bold steps necessary to achieve it.

Space politics also began to influence domestic policies. Besides being a humbling blow to American pride and confidence, the Soviet satellite radically shifted US missile policy and laid the groundwork for the embryonic US space program. Its tremendous propaganda impact had a substantial effect on domestic policies in the United States as well. Seeing itself not up to the task of beating the Soviets in space, the United States revised academic programs and, for the first time, appointed a presidential science advisor. Perhaps most enduring was the passage in 1958 of the National Defense Education Act (NDEA). As a direct response to perceived US educational and technological deficiencies that *Sputnik* represented, NDEA authorized US$188 million (US$1.4 billion in 2008 dollars) to promote study in areas where the United States was deficient. The NDEA provided the funds for a greater number of American college students to undertake math, science, foreign languages, and area studies, which were all deemed as necessary to close the knowledge gap that had put the US behind the Soviet Union. The NDEA also helped to increase the percentage of the US population who attended college from 15 percent in 1940 to 40 percent in 1970.[74]

Additionally, *Sputnik* helped to influence the modern US university system through the broadening of the university-government research partnership that is prevalent today in the United States. Though this relationship had begun during World War II, the Soviet space challenge prodded the federal government into ramping up subsidies for US university overhead expenses, and universities were converted into research tools for national defense. The United States investment in basic research and development increased from US$2.7 billion (US$22.8 billion in 2010 dollars) in 1955 to over US$15 billion (US$120 billion in 2010 dollars) in the early 1960s, with research universities receiving at least one billion dollars annually during the decade.[75] This model would become the standard for many rising space actors, many of whose space programs began as university research programs.

In the end, while Eisenhower accepted the necessity of an institutionalized, national space policy, he did so very reluctantly. Following a 6 March 1958 meeting of the National Security Council, in which Eisenhower was briefed on the requirements for a manned moon program, he wrote a summary in which he expressed his persistent doubt that "there was nothing of value to national security [in space]."[76] But then-Senator Lyndon B. Johnson outlined a US space program's strategic value to national security, explaining that having a civilian program put the first man into orbit would help to further strengthen the legal right of overflight for later military reconnaissance satellites.[77] Eisenhower finally relented.

Eisenhower's about-face can also be attributed to the rapid succession of stunning accomplishments by the Soviets following *Sputnik*. Only a month after *Sputnik*, its sister craft, *Sputnik 2*, achieved orbit, this time carrying a canine cosmonaut, Laika, which was intended as a test of launch survivability for future human space flight. The American public and press's reaction to the launches was a combination of amazement, awe, and fear. What was seemingly at stake was at once tangible and intangible—Americans were left with chilling visions of nuclear bombs raining down on them from orbit. Equally important was the unresolved question as to whether Soviet successes in space had caused irreparable damage to US political influence and prestige around the world, most importantly in the developing world, where so much of the Cold War proxy competition was taking place.

So, despite his reservations, Eisenhower's policies laid the foundation for the creation of the unified, institutionalized space program in the United States. Arguing under the age-old pretext of American exceptionalism that "space was about spying, not because the United States was aggressive, but because the USSR was secretive," Eisenhower crafted a policy that was both subtle and bold, in which the triad of security, peace, and cooperation in space became the official US policy position.[78] This tact was a constant that "stemmed from traditional idealism and respect for the rule of law on the one hand and from Cold War competition for prestige on the other."[79] The contrasting classical impulses of international relations were firmly entrenched in the politics of the space race.

The new paradigm of space policy

While Eisenhower strongly believed that weapon-free space was in the best interest of the United States, he was not immune to the internal politics that drove weapons development and acquisition. Pushed along by strategic initiatives, the US military—the Air Force, in particular—quickly developed plans for space that differed considerably from those of most civilian political leaders. The Air Force advocated plans for what became known as "dual-use" programs, which were ostensibly non-lethal in nature but whose technology could be readily applied to military hardware of high lethality. It is most ironic that Eisenhower, the president who would so strongly issue a very public warning against the "military-industrial complex" in his farewell address, would be the one who authorized the creation of what would become one of the largest conduits and consumers of that growing security aspect of US space policy: the National Aeronautics and Space Administration or NASA.

The successive Soviet achievements in technology, rocketry, and early space flight during the late 1950s sent tremors through the US policy community, not only because the Soviets had beaten the US into orbit, but because of the strategic imbalance that *Sputnik* and its successors signified. *Sputnik*, in particular, represented a new threat to the United States and the West. The Soviet rocket that put *Sputnik* into orbit could also launch a nuclear weapon at the United States, and as with the V-2 before, there was no defense against such a weapon. The geographic buffer of two large oceans that had for almost 200 years protected the United States from direct foreign threats had been rendered meaningless. Thus *Sputnik* represented a substantial deterioration in the national security of the United States, which had since Hiroshima and Nagasaki been based upon nuclear deterrence. Until this point, the US had held the technological, and by extension political, advantage in the Cold War geopolitical tug-of-war. But after *Sputnik*, political and scientific leaders in the US openly questioned this assumption of superiority. The fear that the USSR was technologically ahead of the US led to nothing short of panic in the American policy community, which forced Eisenhower's hand in creating NASA.

Before *Sputnik*, some of the most important assumptions of US foreign policy toward the Soviet Union were: 1) that the US could theoretically win a nuclear war with the Soviets; 2) that the US could not be matched in technology, military, and educational leadership; 3) that the Western capitalist free market was best at creating technological advances in science and technology; and 4) that the US would naturally be the first to launch an artificial satellite because, at its heart, the USSR was a technologically backward and poor agrarian country. These assumptions, as well as working theories and the political strategies that had governed East–West relations since 1945, vaporized along with *Sputnik*'s rocket plume.

But, in the end, *Sputnik*'s most enduring effect was its death-blow to the West's naïve conceptions of space policy, not the satellite's short-term practical value as a prestige weapon. Specifically, *Sputnik* served the same function as the Japanese attack on Pearl Harbor had done 16 years prior— it was the catalyst for a fundamental paradigm shift in thinking about national security policy. Just as Pearl Harbor and the US entrance into World War II marked the beginning of a process of developing a much larger and eventually permanent military establishment in the US, *Sputnik* similarly signaled the beginning of the permanent inclusion of space policy in the policymaking of all space-capable states. The aftermath of *Sputnik* touched virtually every aspect of international relations, although some, even in the Soviet leadership, did not at the time appreciate fully the gravity of the change.[80] A unified US space policy necessarily and quickly emerged because the Soviets had proven themselves equals in technology and education, and the symbolic power of the USSR's technology orbiting out of reach of US defenses seemingly meant that the ability of the United States to win a nuclear war against the Soviet Union had vanished. Of equal importance was the fact that the US failure to be the first in space was perceived as a major public relations debacle in US efforts to entice developing countries to ally with the West.

In the end, *Sputnik* was a godsend to Washington, to the budding US space program, and to the international security of the time. The United States' sense of superiority was shattered and the Soviet Union was cured of its inferiority complex. Seemingly more evenly matched, both sides were forced to find common ground. At least for the short term, despite the continuing Cold War standoff, the possibility of a real "hot" war had become less likely. The then-evolving concept of Mutually Assured Destruction (MAD) was suddenly elevated to a new level. The same fear that drove nuclear brinksmanship during the early 1960s also fed the imaginations of US policymakers in considering ways to attain military and political advantages via the use of space. Perhaps the most imaginative of these ideas was Project Horizon, a 1959 US Army proposal penned by von Braun that would have established a manned military base on the moon by 1966 (the Air Force had a similar plan). The Army later passed the proposal on to NASA, which ultimately shelved the plan.[81]

Spying from space

One of the greatest developments and least appreciated results of the space race was the urgent drive to develop space-based surveillance platforms to perform what is now called remote sensing, reconnaissance, or surveillance. These "spy satellites" would open up a strategic knowledge base never before available, for the first time giving states an unobstructed view of anywhere on the globe. Whether for military or civilian applications, space-based surveillance and remote sensing remains a most sought-after element in national space programs today. It is arguably the most important and durable legacy of this pivotal period in the development of space as an arena of international relations.

Upon hearing the news of *Sputnik*, von Braun's reaction was direct: "I'll be damned," seemingly reflecting his sense of failing to convince Washington of the Soviets' space ambitions.[82] He and others in the US scientific community were understandably concerned about the emerging synergy of capabilities the Soviet Union was displaying in all areas of strategic interest: nuclear weapons, ballistic missiles, high technology, and precision manufacturing. As it turns out, *Sputnik* (and its sister *Sputnik 2*) was merely the tip of the iceberg in Soviet designs for space. Though initially designed for surveillance (as were the later Soviet space stations), *Sputnik 3* was the first of many military-oriented space platforms, and later Soviet plans included a space-based missile system, manned space battle stations, and orbital ballistic missile defenses.

Not coincidentally, the US had the same sort of exceptional plans, some of which actually came to fruition. The Cold War airborne spy programs run by the United States became the stuff of technological achievement and political intrigue. While the United States' U-2 and SR-71 Blackbird spy planes gained a certain notoriety because of their speed and high operational ceilings (Mach 3.2 at 25,000 meters for the SR-71), the goal to spy from outer space became a strategic objective that took precedence over most other space-oriented endeavors, though the inherent secrecy of such projects obscured their public prominence at the time. Following on from NSC 5520, Eisenhower authorized the then-secret space effort by the CIA, code-named Corona, which was the US military's first spy satellite program to become operational.[83] The launch of the first Corona satellite in June 1959 reflected the necessity of space-based surveillance and spurred much of the subsequent development in space programs and related technologies, which became an integral, if not vital, strategic option exercised by the Cold War rivals. This first major US surveillance program, which allowed the United States to photograph the Soviet Union from space, was the later Apollo program's rival in difficulty and achievement. Though the Apollo program would capture headlines around the world in the 1960s, information-gathering systems like Corona were deemed of equal, if not greater, value by US policymakers, particularly President Johnson.

Given the public name Discoverer and using a cover story of space biology experiments, the Corona project was mostly a response to the fear of nuclear attack by the Soviet Union. American leaders faced the pressing dilemma of not knowing what the Soviets were actually doing behind the Iron Curtain, which Corona was meant to answer. Though its photographic resolution was not as good as that of the U-2 spy plane (which was about 30 centimeters), Corona offered a host of other advantages, such as infrared red sensors that were impervious to bad weather, and a location that was out of reach of the Soviet Union's ever-improving missile defenses.[84]

But the covert nature of the Corona project meant that its role in national space policy was generally overlooked and its successes were largely unknown. To this day, space surveillance and information-gathering systems largely remain covert but comprise a substantial proportion of launches by the world's largest space actors. The US and Soviet/ Russian militaries alone launched almost 3,000 military reconnaissance satellites from 1960 to 2010. This trend continues unabated, with up-and-coming space powers such as Brazil, China, and India all having placed surveillance satellites in orbit. Beginning in 1959 with the then-breakthrough image resolution of 7.5 meters (later improved to 1.8 meters), the 144 Corona reconnaissance satellite missions were key to lifting the veil of secrecy from the communist bloc, revealing that the so-called missile gap did not exist (though this fact was held in secret from Congress so as to promote spending on greater missile technology).[85] From the first successful launch and retrieval of film in 1960 to its retirement in 1972, Corona was the United States' indispensible eye in the sky.

While the Soviet Union had led the space race in ballistic missile technology, it trailed in surveillance systems. But not to be outdone, the USSR launched its own surveillance satellite, *Cosmos*, on 26 April 1962, and like the US, the USSR concocted scientific cover stories to mask the satellite's true nature.[86] Cleverly, the Soviets ended up calling *all* their satellites *Cosmos* to hide their true intentions. These operations ranged from spy systems to oceanographic mapping to deep-space probes. By the end of the program, the USSR had launched 2,400 satellites under the Cosmos name.[87]

Other fanciful projects, concurrent and subsequent, reflect the extent to which the race to create space-borne surveillance systems accelerated during this time period and demanded an ever-increasing proportion of space programs' budgets. A classic example was the X-20 Dyna-Soar. Begun in 1957 and based on World War II German feasibility studies of hypersonic bombers as well as the work of the Chinese-born founder of California's Jet Propulsion Laboratory, Qian Xuesen (see Chapter 3), the X-20 was to be a Mach 18 space plane for reconnaissance, bombing ground targets, and destroying enemy satellites. Eisenhower refused to fund the project, but to circumvent this inconvenience the US Air Force reclassified the X-20 as a research platform, which then opened up access to funds.[88] Astronauts were secretly chosen for the program (including the future first man on the moon, Neil Armstrong) and construction proceeded unimpeded until the project was suddenly cancelled in 1963 by Secretary of Defense Robert McNamara, despite the US having already spent over US$660 million on it (US$4.65 billion in 2010 dollars).[89]

On the same day the X-20 was cancelled, the Air Force announced a new space project. Inspired to overcome the vulnerability exposed by the earlier downing in 1960 of Francis Gary Powers's U-2 spy plane over the Soviet Union, the Manned Orbital Laboratory (MOL) was proposed.[90] The MOL was planned to be an orbital spy platform manned by Air Force personnel, but was finally cancelled in 1969. The Soviets responded with their own manned military space stations, code-named Almaz, which operated during the 1970s.[91]

The intersection of purpose

The institutionalization of US space policy finally arrived with the creation of the United States' first civilian space agency. After several months of debate, Eisenhower's cabinet produced a document that declared that a new federal agency was needed to conduct all non-military activity in space. Accordingly, in April 1958 Eisenhower proposed to Congress the creation of the National Aeronautics and Space Agency (NASA), which would use the then-current National Advisory Committee for Aeronautics as its foundation, as well as its more than 8,000 employees and US$100 million in funding. In July 1958, the National Aeronautics and Space Act (H.R. 12575) was enacted as the first legislation that institutionalized space policy in the United States under the NASA umbrella.

Though generally thought of as an organization dedicated to the exploration of space and directed by legislation to be civilian in nature, from its inception NASA has played an integral role as a national-security-oriented space program in which defense-related projects and their foreign policy effects were the driving force. Within H.R. 12575 there is very specific language that affiliates NASA with the Department of Defense: "The Administration [NASA] shall be considered a defense agency of the United States," and states the necessity for NASA officials to "act in harmony with the agencies in the Executive Branch," in particular, the Department of Defense.[92] In other words, DoD had the final word on NASA's activities.

NASA's responsibilities to the national security agenda grew quickly. Eisenhower specifically directed that the military components of fundamental rocketry or space research, such as the Army's Jet Propulsion Laboratory at the California Institute of Technology and von Braun's rocket team at the Redstone Arsenal in Huntsville, Alabama, be put under NASA's control.[93] This approach has various explanations. First, the creation of a civilian space agency was directly in response to the Soviets' more obviously military-driven program, and was perceived to be the best way to build international goodwill within the framework of the ideologically riven Cold War. Second, this decision reflected Eisenhower's overriding fear of the overt militarization of space and his wish for a clear demarcation of chiefly military space research and purely scientific space research.[94] Lastly, it solved a sensitive issue, Redstone being viewed in some quarters as the "Nazi rocket" because of its architect.

Accordingly, the preamble to NASA's charter declares that the organization is dedicated to the "general welfare and security" of the United States but that its purpose would be non-military in nature. This obvious contravention to the purpose laid out by the Space Act had a purpose. On paper, NASA was to function as a civilian space program and the DoD would coordinate its own military space program. Therefore, between 1958 and 1961, in line with Eisenhower's directive, the US Army transferred its Vanguard ballistic missile and space program to the incipient civilian space agency, though not willingly (in part to prevent the Air Force from obtaining it).[95] NASA also inherited military missile programs, facilities (such as the current Marshall Space Flight Center), and the programs that were developing the *Saturn V* launcher for the moon project. Nonetheless, as already described, the development of NASA—as well as other countries' subsequent national space agencies—conformed in many other ways to the national security impetus of space programs. NASA's founding demonstrates how early space policy in the United States became concretized as a part of the national security agenda. This national security remnant is evidenced in a number of other ways.

During the first 15 years of the US space program—from John Glenn's brief three orbits of the Earth to the Apollo program—all US astronauts were either former or active-duty military. The sole exception was made for the *Apollo 17*, for which Dr. Harrison Schmitt, a civilian geologist, was a crew member. Even the director of the Apollo program, Samuel Phillips, was an Air Force general. NASA contracted with the US military and military contractors to supply the rockets necessary to fulfill its objective of launching civilian satellites and spacecraft. Also, until the deployment of the US space shuttle in 1981, all of NASA's primary launch vehicles—Redstone, Juno, Delta, Atlas, Titan, and Saturn—were adapted directly from military rocket programs. For example, the Redstone rockets that eventually became the backbone of the Mercury program were simply transfers from the previous Army Ballistic Missile Agency (themselves direct descendents of the earlier German V-2).[96] Even the *Saturn V*, which sent *Apollo 11* to the moon, utilized technology that was developed from earlier military ballistic missile programs.

This arrangement allowed the DoD to play an important supporting role in NASA's scientific programs. Besides its own outer-space activities, NASA was occasionally recruited to assume other national security responsibilities, as was the case when it provided the cover story for the downing of Francis Gary Powers's U-2 spy plane over the Soviet Union in 1960. NASA publicly claimed that the downed aircraft was a NASA weather research aircraft that had strayed off-course (a bogus story that was exposed shortly thereafter). Even the later space shuttle was conceived and designed to accommodate various military cargoes as well as the requirements of both the US and Soviet militaries. The US Air Force explicitly required that the shuttle be able to put spy satellites into polar orbit.[97]

The same was true of the Soviet Union's space shuttle *Buran* ("blizzard" in Russian), which was built in an effort to maintain the strategic parity between the superpowers. *Buran* was conceived as a response to the US shuttle as well as the Reagan administration's Strategic Defense Initiative, and was intended to support planned space-based military platforms. By the time *Buran* became operational in 1988, however, strategic arms reduction treaties and the imminent end of the Cold War made the Soviet shuttle superfluous, leading to the program's cancellation in 1993.

The official US space policy during the George W. Bush administration, as outlined in *Presidential Directive 5 on US Space Transportation Policy* (2006), codified the primacy of the military role in developing future space policy by requiring all civilian launch vehicles to adhere to military specifications, thus sustaining the dual-use criterion.[98] For all other major space programs, national security considerations—most frequently manifested via military projects—have been the catalyst for the development of launch vehicles and satellite systems. NASA frequently has been utilized by the military to present the "softer side" of the US space program to the public, but has maintained a profoundly close working relationship with the US military, which continues to this day.[99]

The precursors to the Apollo program were similarly security-oriented. It is well known that NASA's Gemini program followed the Mercury manned orbital missions. But it is lesser known that a concurrent program called Blue Gemini existed, in which the Air Force sought to recruit NASA technology for military missions to a proposed military space station (called Manned Orbital Development System—MODS).[100] At first, NASA was receptive to sharing launch costs in exchange for allowing Air Force officers to fly as co-pilots, especially when the military offered NASA as much as US$100 million in compensation. But once the full details of the program became clear, NASA shied away from the idea of using astronauts as high-flying military observers. Ultimately, US State Department reservations about turning NASA into an overtly military space agency led to the cancellation of the program.

A similar story held true for the Ranger and Pioneer spacecraft programs. Announced by Secretary of Defense Neil McElroy in March 1958 in response to *Sputnik*, these probes were meant to "one-up" the Soviet Union by obtaining in-depth information about the moon's surface for future human landings. Responsibilities for the flights were divided between the Air Force and the Army. The NASA/Army *Pioneer 3* probe was the first to fly past the moon at a leisurely distance of 60,000 kilometers. The Ranger program, run by NASA and the Air Force from 1961 to 1965, launched nine probes built by RCA (Radio Corporation of America) atop Atlas ballistic missiles built by Space Technology Laboratories, a private contractor working for the Air Force.[101] The first six exploded after launch, missed the moon entirely, or suffered mechanical failure. Only the last three (7 through 9) were considered successful, sending back over 17,000 photos of the moon's surface.

Race to the moon

There is a long history of government policies, supported by lofty rhetoric and visionary goals, having more down-to-earth strategic motivations. Two well-known examples are Lincoln's Emancipation Proclamation and Eisenhower's Federal Aid Highway Act of 1956, both of which have been traditionally lauded as farsighted policy for the public good when, in fact, they were really stratagems to achieve a national security goal (Lincoln's was to prevent Britain's possible recognition of the South during the American Civil War, and Eisenhower's was to ensure rapid transit of the military during the Cold War).

After the Kennedy administration took office, it was almost immediately confronted with the geopolitics of the Cold War again serving as the reactionary catalyst for the next phase in the US space policy. Collectively, the space policy of the previous Eisenhower administration had been focused on reducing the possibility of a Soviet nuclear attack on the US through photo surveillance and the establishment of the freedom-of-overflight legal regime.[102] Kennedy's space policy built upon these foundations and added human space flight as a priority. Though his science advisors urged that he publicize the space program to promote the cultural, public service, and military importance of space activities, the space program was deemed important by the administration not so much for its scientific value but because it was reasoned to be a superior tool within the scope of the Cold War in gaining international prestige and re-establishing the US position as a world leader in the eyes of the "free world" as well as of developing countries.

Such was the case with President Kennedy's challenge for landing an American on the moon "before [the] decade is out." But more importantly, it was the combination of geopolitical fears and Cold War competition that propelled the first human voyage to the moon, not romantic visions of exploration or even scientific curiosity. Publicly praised for its ambitious and farsighted goal, the moon program, too, was deeply intertwined with, and in fact wholly justified by, the overriding national security objectives of the time.

By 1961 the Soviet Union was demonstrably ahead of the US in terms of most space accomplishments, if not space technology. Facing an apparent Soviet juggernaut of success in space activities, US missile and space policy became a key election issue for John F. Kennedy's 1960 presidential election campaign. The perception that the Soviets were winning the space race was an image that Kennedy exploited in his bid for the White House.

Kennedy accused the Eisenhower administration of creating a "missile gap" with the Soviets by placing fiscal policy ahead of national security.[103] This was an apprehension that Eisenhower's Science Advisory Committee had itself projected in 1957, arguing that the Soviet "military effort" would soon outpace the United States.[104] Soviet space efforts continued to push US space policy on a reactionary course. Though no numerical missile gap existed (which Secretary of Defense McNamara finally admitted in 1961), the incoming Kennedy administration sought a way to propel the United States back to the position of leadership by recreating the previous commanding lead in rocket and space technology that the United States had enjoyed immediately after World War II. The Kennedy administration's solution was to raise the stakes in the space race—which, in this case, was around 384,000 kilometers farther out than recent Soviet milestones.

The new administration's emphasis on space was prodded by the fact that only two years prior the USSR had extended its lead in the space race. Two Soviet probes were the first to go to the moon. One probe, *Luna 2*, had hard-landed (i.e., crashed) on the moon, and another, *Luna 3*, had snapped the first image of the far side of the moon. The US had twice failed to counter the *Sputnik* shock, only managing in February 1958 to finally put the small *Explorer 1* satellite into orbit. Even with this small victory, Kennedy's new space policy was predicated on the fact that the Soviet launchers, like the Americans', were converted intercontinental ballistic missiles, which could just have easily carried nuclear warheads. On 12 April 1961, the Soviet Union's *Vostok 1* carried the first human, Yuri Gagarin, into space. With the bitter taste of *Sputnik* still fresh, Gagarin's flight—along with the humiliating defeat a week later of the US-supported Bay of Pigs invasion of Cuba—forced the Kennedy administration to finally push the US entirely into the space age, thrusting policy changes upon sometimes unenthusiastic actors.

Kennedy became convinced of the merit of a lunar program after consulting with the CIA and NASA, and again with space advocate Vice President Johnson. Kennedy was also swayed by intelligence estimates that a space race was the most advantageous route for the US in this area because neither side yet had a powerful enough rocket for a manned lunar landing, and therefore the odds were even.[105] In essence, like *Sputnik*, Gagarin's flight had again frightened US policymakers into action because, at the time, the United States was not even remotely close to being able to match Soviet achievements in space. The space race entered its final phase with each country feverishly seeking to beat the other to the moon. US intelligence services knew the Soviets had their own moon landing program, but only glimpses and bits of hearsay were available at the time.

In what would be the turning point of the space race, Kennedy, speaking to a Joint Session of Congress on 25 May 1961, announced the goal to land a man on the moon within the decade. Though the Apollo program had been initially proposed by the Eisenhower administration, the new urgency of the Cold War, together with Soviet successes in space, made such bold action seem necessary, if not imperative. In late 1961, NASA entered into discussions with the Air Force's Ballistic Missiles Division to use the Air Force's Titan 2 booster for the new Gemini spacecraft. The deal was mutually beneficial, since NASA's new capsule could become the spy platform that the Air Force had coveted for almost four years. Conversely, NASA officials supported the increased Air Force reliance on Gemini, since they thought it would result in cheaper operations for their own version of Gemini.

The breaking point seems to have been when Secretary of Defense McNamara proposed, for reasons of cost-savings, that the DoD completely assume control over the Gemini program as well as all manned space flights in low Earth orbit (LEO).[106] The Cuban Missile Crisis in October 1962 solidified in American public opinion the absolute necessity of beating the Soviets to the moon and, therefore, the program became vital to Kennedy's political future as well. Having come face-to-face with possible nuclear annihilation, Kennedy recognized the moon program for what it was—a political tool and ploy of paramount importance. All told, the Air Force proposed thirteen new space programs by January 1963, none of which would come to fruition, and the US Air Force never achieved a military manned spacecraft. Regardless of these failures, the US DoD space budget by 1963 already exceeded US$1.5 billion (US$10.5 billion in 2010 dollars).[107]

Domestically, Kennedy was faced not only with public reservations about his ambitious plans, but also reservations within NASA itself. Ironically, Kennedy had to convince NASA's own administrator, James Webb, to support the program, saying "this [the Apollo program] is important for political reasons, international political reasons. This is, whether we like it or not, an intensive race [with the Soviets]."[108] Webb responded that NASA scientists (and some US congressional representatives) still doubted the viability of a moon landing, since practically nothing scientific was understood at that time about the lunar surface (some thought the surface was so soft that the lander would sink). Moreover, Webb argued that "all" of the civilian space program "can be directly or indirectly militarily useful."[109] Abandoning Eisenhower's cautiousness, Kennedy willingly used the apparent desperation of the situation to goad the US into action, as he was convinced that the moon project needed a military rationale. The space race was not to be only a demonstration of scientific prowess; it was to be cast in the setting of a classic battle of freedom versus tyranny, which the United States had to win to preserve its hegemonic role as the leader of the Western world.

Therefore, Kennedy tried a different tactic to blunt the Soviet advantage. Despite the overtly realist rationale behind the early US space program's founding and structure, Kennedy, still unbeknownst to most, tried to also use the moon program as a tool of détente and cooperation with the Soviet Union. With the moon program already underway, Kennedy instructed NASA director Webb to initiate a program for a joint US–USSR space program, including "cooperation in lunar landing programs."[110] The timing was advantageous since the Soviet Union had just recently had its first major setback when its *Cosmos 21* Mars lander failed to escape Earth orbit.

During a 20 September 1963 address to the UN General Assembly, Kennedy made an astounding public offer: that the United States and the Soviet Union pursue a joint moon program. This seeming drastic shift in Kennedy's attitude, from initially throwing down the gauntlet to offering an olive branch, took the Soviets, and many Americans, by surprise. But as a political and economic strategy, it made sense. The two superpowers had come perilously close to nuclear Armageddon the previous year during the Cuban Missile Crisis, which undoubtedly imparted a great sense of caution and practicality on both sides. In addition, the recently signed Limited Test Ban Treaty (LTBT) had created some rapprochement between the superpowers.

As it turns out, this very public effort at détente was merely part of a larger scheme in which Kennedy was attempting to reshuffle the deck of political support for the space program in the United States as well as to redefine US-Soviet relations. Kennedy had privately made the same offer to Soviet Premier Khrushchev months before, and Khrushchev at first declined for fear that the US might learn Soviet technological secrets (the Russians were, at the time, ahead in heavy launch vehicles).[111] Kennedy then reiterated the offer to the Soviet ambassador only a month after the UN speech. These overtures might have seemed peculiar, but they reflected a partial shifting sentiment toward space and a perceived need to share what was shaping up to be the seemingly overwhelming financial and scientific burden of space exploration, especially for the Soviet Union, which was strained under the cost of the Cold War.[112]

Public opinion played a large role in the decision as well. Domestic support for the space program in the United States was declining. By early 1963 there was a mounting chorus of public criticism over the cost of the program, which critics charged was untenable. The Apollo program was to reach a price tag of over US$25 billion (US$176 billion in 2010 dollars), accounting for nearly 4.5 percent of the federal budget.[113] Kennedy asked Vice President Johnson, in his role as head of the National Aeronautics and Space Council, to review the program, thus virtually guaranteeing a positive response. But Kennedy's bold political gamble with Apollo persisted. Indeed, it seems that the gambit might have borne fruit, as Khrushchev had apparently changed his mind about cooperation, but his change of heart was made moot just 10 days later by an assassin's bullet in Dallas, Texas.[114]

After Kennedy's death, President Johnson continued the push for space exploration and to expand the space program's parameters. While still emphasizing the symbolic role the space race played in the Cold War, Johnson broadened the scope of US space policy to include the importance of technology, science, commercial applications, economic stimulus, military applications, and of course, its political benefits.[115] Thus, while continuing to support the US space program with the same enthusiasm as Kennedy, the Johnson administration was the first to fully appreciate publicly and use privately the wide spectrum of policy areas that were affected by space activities; in other words, to really understand and implement space policy in the modern sense.

The other moon program

Though modern Western history is replete with stories of the saga of the *Apollo 11* mission to the moon, less is generally known about the Soviet Union's parallel effort to reach the moon ahead of the United States and the consequences this effort had for space policy, then and now. The USSR had, in fact, two massive secret projects designed to win the space race, and for a time, unbeknownst to most, the space race was indeed a spirited competition. The fact that the Soviets were engaged in their own program was known to American intelligence, but its size, capabilities, and organization were not. A then-classified CIA intelligence estimate calculated that the Soviet Union had allocated US$7 billion (US$50 billion in 2010 dollars) for its space program and revealed that the Soviets were building massively large booster rockets whose purpose could only be for achieving lunar orbit.[116] These estimates of Soviet efforts were leaked to NASA officials, which caused the *Apollo 11* landing timetable to be accelerated by at least one year.[117]

In the post-Cold War period, declassified intelligence estimates have revealed one of the great ironies of the space race. In the Soviet Union, a society that abounded in the centralization of power and authority, the moon program was, in fact, disjointed, fragmented, and very poorly managed, and thus was incapable of promoting interdivisional cooperation—precisely what would have been necessary to beat the United States. As it turns out, a key choice the USSR faced was one that many states have faced since the dawn of the missile/space age, which was whether to invest in the more immediate returns, represented by ballistic missiles, or the longer-term returns of a space program. As it turns out, the Soviet military was more interested in strategic ballistic missiles than a moon landing, and was doubly indisposed to support what was clearly a political program that yielded little military benefit.

Thus, the Soviet moon program was hampered, if not doomed, from the beginning by a Soviet space culture that contained "mutually antagonistic entities with an ad hoc pattern of shifting alliances and animosities" and whose typical decision-making process was, at best, arbitrary.[118] This systemic handicap was exacerbated by intense interdivisional military rivalries that operated on a shoestring budget. Ironically then, though the United States was ideologically and structurally the antipode of the USSR in so many ways, the US emphasis on central planning and interagency cooperation became the key to the success of the Apollo program.

Under the series called Zond ("probe" in Russian), the first step in the Soviet program, called the L1 project, would send a Soviet crew around the moon before the US, using a stripped-down Soyuz spacecraft launched by newly designed N-1 booster rocket (the Soviet equivalent of the *Saturn V*). The Soviet government approved production of a two-phase program in 1966, beginning with the production of 19 lunar-orbiting vehicles. A sister program, the L3 project, was intended to beat the Apollo program to the lunar surface.[119] Designed by Sergei Korolev, the Soviets' chief space engineer and rocket designer, the programs were developed in the early 1960s and had a timeline that called for the first Soviet moon landing in September 1968. These programs were highly ambitious and extremely complex, involving the dozens of launches for the in-orbit construction of a multi-stage lunar vehicle.

Reportedly not having paid much attention to Kennedy's moon-landing pronouncement, Khrushchev finally sanctioned the Soviet moon program in 1964 as part of his "missiles-first" policy. He reasoned that a moon program would benefit the development and improvement of military ballistic missiles, a belief that had become especially acute following the Cuban Missile Crisis.[120] The Soviet moon program was intended to continue the established one-upsmanship in space by defeating the Americans at their own game. While Khrushchev was ousted from power that same year, construction nonetheless continued under the new Brezhnev regime. The drive to beat the Americans continued unabated, at least for a while.

Like a skilled boxer, the Soviets quickly landed several well-placed jabs in the space competition, blunting the United States' best efforts. Valentina Tereshkova became the first woman in space in June 1963; then Alexei Leonov became the first man to conduct a spacewalk in March 1965, three months before the US accomplished the same feat. The next year *Luna 9* became the first spacecraft to soft land on the moon, from where it transmitted the first pictures from the lunar surface. Its sister craft, *Luna 10*, became the first lunar orbital probe just three months later. For a period in the mid-1960s, the Soviets commanded the high ground of space and had put the United States in the position that Eisenhower had dreaded—playing catch-up. Just as it seemed the Soviet space program was gaining sufficient momentum to permanently overtake the United States, Sergei Korolev died unexpectedly in 1966.

The CIA learned about the heavy rocket being constructed for the Soviet mission.[121] This intelligence prompted NASA to prepare *Apollo 8* for a possible lunar orbital mission in 1968, ahead of schedule. This acceleration was seemingly warranted by photos taken by a Corona spy satellite, which revealed substantial unexplained activity at the Baikonur

Cosmodrome throughout December 1968, specifically in the appearance of a Proton booster rocket. It then mysteriously disappeared just a few days later, prompting speculation that mechanical problems prevented the Soviet Union from upstaging *Apollo 8*.[122]

Undeterred, the USSR tried again in July 1969, this time to beat *Apollo 11* in landing on the moon, and again the N-1 booster rocket failed in a catastrophic explosion on the pad. While it was apparent to most observers that the United States was about to accomplish Kennedy's goal of landing a man on the moon, the Soviet Union doggedly continued to try to salvage some prestige from their efforts. On 13 July 1969, only three days before the historic launch of *Apollo 11*, the Soviets' *Luna 15* unmanned probe was launched from the Baikonur Cosmodrome in Kazakhstan, hoping to soft-land on the moon just as *Apollo 11* was leaving the Earth, thereby stealing some of the Americans' glory. Though it beat *Apollo 11* into lunar orbit by two days, *Luna 15* ultimately failed and crashed in obscurity on the moon's surface just one day after Neil Armstrong's celebrated "one small step." Though defeated in the race to the moon, the USSR was determined to be recognized as a space power, and immediately forged ahead. Some plans were audacious, such as the proposed Mars landing by 1971.[123] Finally, financial constraints turned Soviet attention to near-Earth activities, such as constructing space stations, an undertaking in which the USSR excelled. It would be this endeavor that would signal the next phase and broadening of space power.

Thus, in *Luna*'s failure and *Apollo*'s triumph, the chapter of the last great space competition closed and the next era of the space age opened, which would be characterized by both a reduction in tensions over space and the emergence of a new class of divergent space activities that included a broader range of state actors. With the moon landing completed, Soviet capitulation in the space race, and the Vietnam War unresolved, public sentiment toward space in the United States turned sour. In his praise for the recently ratified Outer Space Treaty, US President Lyndon Johnson stated that "This treaty ... reserved an unspoiled area for strictly peaceful purposes to benefit all mankind."[124] In reality, even before the outcome of the moon competition was known, Johnson wanted to find a way to deemphasize space projects, "[defuse] the space race between the US and the Soviets," and create international cost-sharing in space in order to funnel more money into the Vietnam War.[125]

Perhaps following Johnson's lead, the new Nixon administration sought to change the space paradigm toward the type of cooperation that Kennedy had once advocated. The tactic employed by the United States to blunt other states' attempting space programs was to counterintuitively declare space a common area for all humankind. The greatest expression of this cooperation was the Apollo-Soyuz Test Project (typically just called Apollo-Soyuz), which culminated in the orbital docking of US and Soviet spacecraft in July 1975.

This singular act of Cold War détente was, like its space race predecessors, an outgrowth of the international political environment of the time. Both superpowers were interested in Apollo-Soyuz because it served their respective political goals. For the USSR, it helped to mend its bellicose image after the 1968 invasion of Czechoslovakia and border skirmishes with China. For the US, strategic parity with the Soviet Union in the early 1970s meant massive military spending increases would be necessary if the Nixon administration could not reduce tensions with the Soviets.[126]

The contemplated militarization of space

While the moon programs provided the soft power of prestige for the United States (also sought by the Soviet Union), the evolution of space programs was almost exclusively driven by, and therefore focused on, perceived national security concerns. The superpowers' national space programs were, from the beginning, primarily military in nature and composition, and oriented toward providing prestige, intelligence, and communications as well as improved ballistic missiles systems for delivering nuclear payloads. Even satellite systems, today an essential lynchpin in the framework of modern life, were the product of the attempted militarization of outer space. Though generally not having direct offensive capabilities, satellites quickly proved themselves indispensable to the militaries of the larger nation-states in providing intelligence, communication systems, and global positioning information. Today, modern militaries and intelligence agencies are inexorably dependent on space-based information and communication systems.

But while security concerns typically did provide the basis for otherwise overtly military space projects, there have been a few notable and ambitious programs that did attempt the explicit militarization of space.[127] As has been described, the trend to use space research as a façade for defense-related programs emerged parallel with the space race. While in the minority in the overall scheme of national space activities, military space programs have provided a template for a continued physical role for defense-related projects, which not only continues to this day but is being emulated by the larger, aspiring space actors of the developing world.

Though the idea of space weaponization gained popular notoriety with the Reagan administration's Strategic Defense Initiative (popularly known as "Star Wars") during the 1980s, the concept had actually emerged during the Apollo program. As early as 1962, the United States had already tested an anti-satellite system, called Program 505, which used the US Army's Nike Zeus rocket to shoot down enemy ballistic missiles and, potentially, satellites.[128] Its successor, Program 437, used a more powerful Thor rocket that was to disable or destroy upper-atmosphere or space targets either directly with a nuclear explosion or indirectly via the resultant electromagnetic pulse, which destroys electrical systems.[129]

The Soviet Union also developed space-based military programs. After the miscalculation of the Cuban Missile Crisis, the USSR developed various space-based tactical systems. As early as 1960, the Khrushchev approved the Istrebitel Sputnikov ("Destroyer of Satellites" in Russian), which was a co-orbital anti-satellite system that was to be launched into the same orbit as the intended target, approach it, and explode its shrapnel warhead, thus obliterating the target.[130] Another was the so-called Fractional Orbital Bombardment System (FOBS), which would launch a nuclear warhead into a circumpolar orbit where it would approach the United States from the south, thereby circumventing the NORAD radar early warning system (which was oriented toward an attack coming over the North Pole).[131] In these and many other proposed systems, space achieved a level of equal consideration in defense planning.

By the 1970s, space had become so intertwined with the superpowers' strategic outlook that it was even a considered factor in reducing Cold War tensions and was an element in many of the strategic arms limitation talks (SALT) of the 1970s. Besides covering traditional land, sea, and air forces, the agreements emanating from the SALT I and SALT II talks pointedly also included constraints on space-based weapons and bans on their testing.[132] Moreover, verification of the United States' and the Soviet Union's adherence to the agreements were to be conducted, at least in part, by space-based methods (i.e., via spy satellites). These talks and the subsequent agreements set the stage for the post-Cold War era, in which established and rising powers alike have looked to space to bolster security capabilities.[133]

The space programs of other developed countries

While the United States and the Soviet Union were using their respective space programs as Cold War proxy competition, other developed countries also made the decision to dip their toes into the ocean of space. Some states chose to follow the US model by creating separate space programs for civilian and military applications, while others opted for a merger of these two functions. While the stated goals for these other space programs rarely contained the overtly competitive or security-oriented rhetoric of the superpowers during the Cold War, Western European and Japanese programs, among others, nonetheless contained some strong elements of national security as their *raison d'être*, though in each case, the individual trajectory of development was subject to the national (or multinational) realities of the time. It is not surprising that these countries were the first non-superpower space actors, considering that many of the initial efforts of these US allies were made in collaboration with the United States, which sought to ensure that only the "free world" gained access to space; thus the US sponsoring of science and technology in Europe and other countries was founded in political and strategic considerations.[134] As in other events and conflicts of the Cold War, the US sought to exercise its containment policy, though now in the realm of space.

European space programs

In their evolution, European space programs closely resemble those of the US and the USSR, but with important exceptions. While a national security mindset drove the space programs of the Cold War superpowers, such considerations were not initially as prevalent in Western Europe, despite the fact that by the mid-1950s both Britain and France, the two most important postwar space actors in the region, had already committed themselves to the development of independent nuclear and missile forces. Europe's rededication to missile, nuclear, and space programs also had a strong domestic agenda, which was to stem the postwar emigration of top European researchers and technicians so as to not lose Europe's traditional strengths in science and technology, long deemed the keys to national power.

While *Sputnik* did prompt ballistic missile and space-related innovations in Western Europe, it did not generate the same alarm among European policymakers that had gripped their American counterparts.[135] In fact, some European leaders saw *Sputnik* as being a welcome occurrence because it re-emphasized the value of the Western alliance and US vulnerabilities, thus reaffirming the importance of Europe in US foreign policy.[136] The importance of this became more acute with the passage of the Atomic Energy Act (also known as the McMahon Act), a 1946 US law that prohibited the United States from sharing nuclear secrets, even with loyal allies (though Britain was largely exempted).

Nonetheless, initial research by Britain and France into ballistic missile, nuclear, and later space technologies did have its genesis in national security concerns.[137] As the first country to have experienced a sustained ballistic missile attack, Britain was understandably eager to develop its own missile forces. Immediately after the war's end, the Labour Government advocated the creation of a British nuclear deterrent, though this idea was not universally supported in Parliament. To push through the initiative, Prime Minister Atlee established a top-secret Cabinet committee named GEN 75, which outlined the urgent nature of the program, calling it necessary "to save civilization."[138]

Minister of Defense Duncan Sandys (Winston Churchill's son-in-law) oversaw the initial phase of Britain's ballistic missile program.[139] As early as October 1945, Britain was launching captured German V-2 rockets, first from Britain and later from a site in Woomera, South Australia—and following the American lead by also using captured German engineers.[140] By the mid-1950s, at the strong urging of the United States, Britain had developed its own ballistic missile called *Blue Streak*. Based on American *Thor* missiles, Britain's missiles were meant to complement US intermediate-range missiles in Europe.

As in the US, subsequent British advances in rocketry were largely supported and driven by the military, especially in the ostensibly civilian upper-atmosphere sounding rocket program during the 1950s.[141] Like so many military missiles before and after, *Blue Streak* found extended life in 1967 as the first stage of the first European satellite launcher *Europa*.[142] While Britain proposed its own satellite launch program, it never achieved independent launch capability, relying initially on the United States to put its satellites into orbit and then becoming an integral partner with France and Germany to form the nucleus of a collective European space program.

Similarly, for France the motivation for a space program began with matters of defense, but the program evolved to represent French nationalism as well. Its genesis is found in French President Charles de Gaulle's plan for an independent *force de frappe* (strike force) to ensure French territorial integrity would never again be violated. France's nuclear and missile strategy began immediately after the US nuclear attacks on Hiroshima and Nagasaki with the creation of the Commissariat à l'énergie atomique (CEA). Beginning in 1947, a complex with nine launch facilities and various nuclear weapons test sites was constructed in the French colony of Algeria. France's first nuclear test, *Gerboise Bleue*, took place in 1960 at a French military base near Reggane and subsequently led to the development of a bomber and submarine force armed with fission and thermonuclear weapons under the doctrine of *destruction assurée* (assured destruction). This policy simply stated that any nuclear attack on France, whatever its extent, would be meet with a total retaliatory response from French nuclear forces.[143] France also initially focused its efforts on the military applications of ballistic missiles and nuclear weapons, which de Gaulle viewed as tools of national independence from the United States and NATO. Coming of age as a nuclear power later than the US or the USSR, France took a strategic direction less influenced by air power debates and the experience of World War II than had been the case for Britain.[144] France's first indigenous ballistic missile, *Véronique*, lifted off in 1954 from Reggane in the southern Algerian desert. Under de Gaulle's leadership, France eventually developed and implemented a nuclear triad of air, land, and sea-based nuclear deterrence, thus mirroring the arrangement deployed by the United States and the Soviet Union.[145]

These defensive efforts led directly to the establishment of French national space initiatives. After *Sputnik*, the French government established the Société pour l'Étude et la Réalisation d'Engins Balistiques (SEREB) as a civilian space program whose concurrent mission was to further technology and rocketry experience for the French military. In 1962, France established its national space agency, the Centre National d'Études Spatiales (CNES), which oversaw France's becoming the third country with indigenous launch capabilities in 1965 when its first satellite, *Astérix*, was placed in orbit. Like NASA, the CNES is also involved in military space activities, in association with the French Ministry of Defense.[146] Since its emergence as a space power, France has been Europe's most active participant in space projects; it was an instrumental partner in the creation of the European Space Agency (ESA) in 1975 and today hosts the agency's headquarters in Paris. The ESA's launch site, at Kourou in French Guiana, was established by France in 1962.

The contrast between European and American space policies became sharper when Britain and France, within the context of greater European cooperation and the European Economic Community, opted for collaboration in their space programs and came to an agreement to jointly build launchers. This set the stage for a pan-European space program, which occurred at the same time that the United States was caught in the seemingly inescapable vortex of the Cuban Missile Crisis. Having previously been part of the founding of the European Organization for Nuclear Research (CERN) in 1954, the prominent scientists Edoardo Amaldi (Italy) and Pierre Victor Auger (France) took the lead and formally proposed the creation of a European space program. In 1964, two intergovernmental organizations dedicated to the development of scientific satellites and launchers were created: the European Space Research Organization (ESRO) and the European Launcher Development Organization (ELDO).

Then in 1975, ten Western European countries founded the European Space Agency (ESA). In this endeavor, Europe consciously opted for a trajectory different from the US or the USSR in the development of its space program. Instead of focusing on military launchers as the foundation of the program, the ESA instead decided to develop strictly commercial launchers, notably the *Ariane* rocket, which made Europe a world leader in commercial satellite launch services. This difference in *raison d'être* is reflected in the ESA's original charter, which specifically states that the purpose of the Agency shall be to provide for and promote, for exclusively peaceful purposes, cooperation among European states in space research and technology and their space applications, with a view to their being used for scientific purposes and for operational space applications systems.[147] So while the United States and the Soviet Union had utilized their respective space program as tools of competition, the European approach was to consciously use national space programs as an instrument of continued European integration, where "national loyalties can be diffused" creating a pan-European sentiment in which national competition is submerged.[148]

Nonetheless, even in Europe, space policy has evolved along with the changing times. While many other emerging space actors have followed the US example of separate military and civilian space programs, the ESA has slowly been moving toward the integration of both areas under one roof, concurrent with the increased integration of most policy areas within the tutelary realm of the European Union. A 2003 EU Commission White

Paper on European space policy defined outer space as a tool for the implementation of EU policies, including security and defense, and examined potential synergies between military and civil space applications.[149] In 2007, the 25 constituent members of the European Union, along with Norway and Switzerland, agreed on a common space policy, with the aim of creating a common political framework for space activities in Europe. Among the joint goals was the establishment of a common vision and strategy for space exploration as well as for security and defense matters in space.[150] This is in line with the ongoing integration process throughout the EU, which includes the creation of a comprehensive security research program that puts space-related activities at the forefront of guaranteeing European security.

As such, EU defense policies now mirror those of other major space actors in the incorporation of military considerations into space policy. Also similar is the movement of the ESA toward establishing dual-use systems, thus making it more ambiguous where civil and military interests diverge and establishing a joint organization whose future remains to be written. Consequently, the early evolution of European space policy differs sharply from the American experiences in that military and civilian space programs have been merged, in part to foster and protect the autonomy of Europe from friend and foe alike.

In contrast to the European example, other developed states' experiences offer a different template from which to understand the evolution of space policy in the twentieth century. The list of developed states that have developed a space program and achieved some level of experience and sophistication in space-related technology is long and expanding. The most noteworthy example of a non-European country that has followed a different path is Japan, for whom missile and space activities have developed almost wholly in the service of science and economic development.

The Japanese space program

Japan became the first Asian country (and fourth country overall) in space in 1970, having begun its space program in the mid-1950s. Since then, Japan has become one of the most successful space actors, with successful robotic missions to the moon, Mars, and Halley's Comet. But given that it is the only country to have suffered a nuclear attack, Japan's postwar policies regarding nuclear and missile technology have been understandably reserved and strongly influenced by the country's general pacifism, even to the exclusion of the development of military ballistic missiles or weapons of mass destruction that typically formed part of emerging space programs in other countries. Thus, the Japanese space program offers a strong contrast to the American, Soviet, and even Western European examples.

Instead, Japan has focused almost exclusively on developing its ballistic and space technology for peaceful purposes. When Japan's legislature, the Diet, approved the Atomic Energy Basic Law in 1955, it contained a proscription of the development and employment of nuclear weapons. Many times Japanese security experts have weighed the costs and benefits of going nuclear, and most have concluded that such a move would be extremely detrimental to Japan's security interests because it would encourage greater nuclear proliferation in the region,[151] although Japan has built up a relatively powerful military that has incorporated advanced missile technology. This policy decision is based on the simple fact that Japan imports almost 100 percent of its petroleum requirements and the vast majority of its other energy and manufacturing resource needs. Therefore, Japanese policymakers have reasoned that the country's best foreign policy is to maintain a non-confrontational position in international affairs.

Today, Japan is one of the few countries with independent, proven space launch capability. Beginning with experiments run by the Institute of Industrial Science at University of Tokyo, modern rocketry took its first step in Japan in 1955 when, as one of the participant countries in the IGY, it launched a sounding rocket that reached an altitude of 60 kilometers. From this modest beginning, Japan has become an important space actor. Since 1970, Japan has put into orbit nearly 100 satellites and successfully sent up dozens of probes to study celestial objects (including asteroids, comets, the moon, and Mars, with planned missions to Venus and Mercury) and to conduct space science.[152] Though its earlier rocket launchers were hybrids of US and Japanese components, Japan's new H2-B rocket is, by design, entirely Japanese in origin, which is meant to bolster Japan's standing and national pride as well as strengthen its position vis-à-vis the European Union, Russia, and the United States in the international launch-services market.[153]

In 2003 Japan unified its space and missile programs under the Japan Aerospace Exploration Agency (JAXA), which absorbed three previously independent space and rocket agencies (the National Space Development Agency, the Institute of Space and Astronautical Science, and the National Aerospace Laboratory of Japan). Most recently, faced with greater regional instability as evidenced by North Korea's 2009 missile test over Japan, the Japanese government actively considered the construction of a land-based anti-ballistic missile system, which was troubling to some because the technological line between missile defense and weapons of mass destruction is thin indeed (in the end, Japan chose a US-built, ship-based system instead).[154]

True to its primarily science-oriented goals, in 2007 Japan sent a scientific probe to the moon (it had sent a crude predecessor in 1990). The *Kaguya* spacecraft orbited the moon for almost two years before it was deliberately crashed into the lunar surface in June 2009. On board were two high-definition video cameras whose main purpose was to demonstrate Japanese technical capabilities, including high-resolution mapping of the lunar surface. In the end, however, even Japan's ultra-pacifist missile and space policies have been swayed to some extent by the pressure of regional security challenges. Following the 3 September 1998 test of a North Korean missile over the main Japanese island of Honshu, Japan initiated an accelerated program of missile defense and the creation of a spy satellite program, which currently consists of five satellites. In 2008, the Japanese Diet passed a law allowing the lifting of Japan's self-imposed restriction on the military use of space for defensive purposes. This would remove the final obstacle to Japan become a major space actor and full-fledged competitor in commercial and state space activities.

A template for the twenty-first century?

The development of ballistic missiles and, consequently, space programs in developed countries after World War II forever altered strategic policy-making in the international system. Strategically and economically, space programs became an integral element of the national policies of almost every developed state that aspired to assure its territorial and economic integrity—even if they did not possess the necessary launch capabilities. The precedent set by these early space actors has essentially dictated the necessity of a space program to address a country's strategic, economic, and developmental needs. That being the case, the evolution of space programs during this formative period was, first and foremost, an exercise in the extension of traditional security concerns taken from the realist paradigm. Strategic concerns drove the development of the earliest programs and continued to command the lion's share of expenditure on space technology. No major aspiring space actor has thus far completely escaped the vortex of national security in formulating its space policy—if it is not present from the beginning, elements of national security eventually become interwoven into the technological and political tapestries of national policymaking.

Nonetheless, the post-Cold War period has witnessed an amplification of the uses of space. From being merely an arena for demonstrating prestige and power, space has acquired an elevated importance in the growth of the modern "information society," the monitoring of weather and global warming, and democratizing the distribution of boundless information via the Internet. This is not to say that defense or power-projection has by any means diminished in importance as a constituent part of states' space policy; it has not, and, by many measures, continues to increase as a growing number of countries pursue the route to utilizing space for their national interests. The remainder of this book examines, categorizes, and details the space programs of developing countries as well as the principal policy areas in which space is being utilized by the developing countries for their national welfare, socioeconomic development, and national security.

3 First tier space actors

Launching BRICS into space

> *Every gun that is made, every warship launched, every rocket fired, signifies in the final sense a theft from those who hunger and are not fed, those who are cold and are not clothed.*
>
> Dwight D. Eisenhower[1]

Since the end of the Cold War, the gap between the ambitions, achievements, and relative power of developed and developing states has begun to narrow in a number of important areas, including economic performance and influence in the international system. And while the space activities of the Cold War period were dominated by the United States and the Soviet Union, the post-Cold War period has seen the space programs of developing countries begin to approach, in some areas, the capabilities of these space superpowers. Since the late 1980s, a few of the most advanced emerging space actors (EMSAs), such as China and India, are able to accomplish many of the basic functions in launch services, satellite construction, and basic space science that were previously the sole domain of the superpowers.

Many other developing countries have embarked on space programs, albeit on a smaller scale, and an increasing number of developing countries, convinced that the hallmark of national space capability is the ability to utilize space-based resources, are now developing launch vehicles, launch sites, and associated support satellite services. Further, a multitude of other developing countries have laid out space policies that are largely cooperative with more capable space actors. Regardless of their demonstrated or aspired capabilities, an increasing number of developing countries now patently embraced space activities as an integral part of their developmental and national security policies.

From luxury to necessity

For essentially all modern states, the trait most associated with a national space capability is the ability to access space and to utilize space in the promotion of both domestic and foreign policies. Since the 1970s, both developed and developing countries have been expanding their investment in multiple space-related areas, including satellites for communications, weather, reconnaissance, and global positioning (GPS) as well as ground-based hardware and software to process the data received. The most widely used satellites are for communications, which provide service for television, cell phones, and various military purposes. At the same time, countries have been expanding their investments in space-based telecommunications, weather monitoring, remote-sensing satellites, ground receiving equipment, and data processing hardware. The trend toward developing the commercial uses of space as well as scientific research into new commercial applications is generating growing interest and financial commitment among the developing countries of the world.

The geographic location of a country and its proximity to other countries is a highly influential, if not crucial, factor in determining the nature of its space policy, particularly in achieving independent launch capability. The first reason has to do with the physics of achieving escape velocity. Simply put, the closer the country is to the Earth's equator, the faster the launch speed available. This is because at the equator the Earth's rotational spin is higher (1,670 kph) than closer to the poles. The farther north one goes, the more power needed to achieve equatorial or geosynchronous orbit, and consequently, the greater the fuel requirement. It is for this reason that United States, the Soviet Union/Russia, and the European Union have located their heavy-lift launch sites as close to the equator as possible, and that rising space actors with indigenous launch capability have done the same. For example, as discussed later in this chapter, the earlier military government in Brazil took great pains to position its Alcântara Launch Center almost on the equator, even going so far as to expropriate land from poor peasants to make it a reality.

Critics of space programs in the developing world typically argue that the money spent on these endeavors could better be spent addressing more pressing terrestrial necessities, such as poverty reduction. These criticisms follow the classic "guns or butter" argument by implying that by investing in space technologies, EMSAs are denying their populations something more immediate, tangible, and beneficial. But, as will be demonstrated, like the space actors that preceded them, EMSAs from across the range of space capabilities perceive the addition of national space programs and/or space-based technologies to be a prudent decision that has not only yielded some important short-term benefits but will ultimately leave them in a much better position economically and strategically in the longer term (e.g., the US had a 17 percent poverty rate in 1969, but proceeded nonetheless with the first moon landing).

As discussed in Chapter 2, the advent of rocketry in the early twentieth century and the subsequent space competition during the Cold War literally forced upon wealthier, technologically capable states the creation of national space policies and the integration of space activities into their broader foreign policymaking. This meant that space-related endeavors promptly became part and parcel of the promotion and/or defense of their national interests. Nonetheless, in many important ways, despite its unlimited tactical and strategic potential, space was and has continued to be largely perceived by policymakers as simply an extension of geopolitics. Accordingly, previous inclinations in international relations, in areas such as interstate relationships and the degree of cooperation between more dominant and less powerful states, have generally been unchanged, even though states have become ever more involved and dependent on space-based assets and even though the number of states utilizing space has increased.[2] In the information age, space has been utterly indispensible to states' national security and contributes significantly to technological strength, which in turn has contributed to states' national security in the classic strategic sense but also from socioeconomic perspectives.

The inauguration of space policy in developing countries occurred later, sometimes much later, than was the case for the Cold War superpowers and their wealthy, developed allies. Nonetheless, the rationale and impetus for spending their comparatively smaller available funds on space programs has been essentially the same, with a strong tendency to follow the same developmental trajectory as that of the more traditional space actors. Given that military, strategic, technological, commercial, and even subjective cultural impulses can and do affect states' decisions to invest in space, these same motivations are naturally present in the space ambitions of many EMSAs. Though the entry cost for many developing states to establish autonomous space programs remains very high, a number of developing countries nonetheless pursue varying types of space activities, even if it is merely through cooperative arrangements with already established space actors or other emerging space actors, or to develop and exploit a niche either not developed or now discarded by the developed space actors (DVSAs). This type of cooperative arrangement occurs because of the decreasing costs of the space technology, which progressively lowers the barriers for participation. It also allows EMSAs to develop specializations that will permit further cooperation with other DVSAs and, increasingly, among the EMSAs themselves, sometimes eschewing DVSAs altogether. But, in the end, the primary goal of the space policies of the EMSAs is to promote the sovereignty and socioeconomic development of the state; as will be discussed, the definition and character of these objectives differ from country to country in terms of their investments and their utilization of space assets.

This *raison d'être* follows other aspects of the foreign policies of emerging space actors. Over half of the space systems currently in existence are used in support of national militaries; therefore, when EMSAs use space assets to promote national security interests, such as expanding prestige and security, it is simply par for the course.[3] This trajectory is predicated on the countless previous experiences of states' entrance into new areas of international relations. Developed states have long pursued technological development that is primarily in reaction to pressures and/or challenges presented by the still largely anarchic nature of the international system. This same assumption can be applied to the emerging area of space programs and related technology development in the developing world.

As Klein (2006) notes, EMSAs engage in the same cost/benefit calculus as developed space-faring states—when the benefits are greater than the cost, the decision to pursue space activities is a relatively easy one.[4] Accordingly, the sophistication of space programs of EMSAs and the integration of these space programs into states' domestic and foreign policies are positively correlated to states' previous history of security issues and the level of development of the local defense industry. Specifically, the evidence to date from the developing countries of Asia and Latin America strongly suggests that developing states which have experienced sustained periods of regional conflict and/or have endured bouts of international pressure are more likely to develop indigenous defense industries.[5] Given that almost every developed space program has emerged, at least initially, from defense-related research and policies, it follows logically that the trend in the space programs of EMSAs will continue to follow this established evolutionary pattern as the wealth of a country allows. As discussed herein, without exception, this has been true for the most economically and technologically capable of the EMSAs.

Beyond traditional security concerns, the inclusion of space policy is increasingly important to many developing states' economic and social development. Science and technology in the developing world are largely socially, politically, and even culturally constructed, just as in the developed world.[6] Even as the most technologically capable EMSAs, such as China, build their own constellations of spy and communication satellites, they are careful not to neglect the very real and tangible advantages to be accrued through space systems, even if dual-use. Thus, while a select few EMSAs have included manned space flight as an objective, for the rest, satellites and the very tangible national security information and socioeconomic benefits they make available are what really define the EMSAs' use of space. In the end, it matters not whether the satellite image shows an approaching hurricane or foreign invader, the ability it affords the satellite's owner to protect the state and its population is the same.

Besides the bipolar nature of the East–West conflict during the Cold War, one of the traditional constraints on the space programs in developing countries has been restrictions placed on the export of space-related technology. Before 1992, all US satellite-related technologies were classified as "munitions" and therefore subject to regulation by the US State Department under a regime known as the International Traffic in Arms Regulations (ITAR). During the mid-1990s, these restrictions were eased for "dual-use" technologies, which are those not exclusively military in purpose and application. The line between the two concepts in practice, however, is nebulous, since essentially all space technology is dual-use.[7]

These coveted space systems have become an integral, even indispensible, part of daily life around the world and are used for myriad purposes. Among the most commonly employed satellite systems are telecommunications, navigation, weather forecasting, Earth observation, and military reconnaissance. Depending on their explicit or public purpose, satellites that look at the Earth are alternately called remote-sensing, reconnaissance, or observation satellites. As early as the 1860s, Jules Verne had written of "lunanauts" who could monitor weather systems. Later balloons and aircraft took photos of the Earth and were able to appreciate its curvature. To paraphrase the old saying regarding beauty, the purpose of remote-sensing satellites is in the eye of the owner.

The world's first remote-sensing satellite was the *Television and Infrared Observation Satellite* (*TIROS-1*), which was launched by the US in 1960. *TIROS-1* and its successors formed the first early weather-satellite systems that allowed the US for the first time to observe global weather. The first civilian remote-sensing satellite, *Landsat*, was launched in 1972. In the beginning, a clear distinction existed between civilian remote-sensing satellites and government imaging satellites, since the civilian technology was far inferior to the technology used for military reconnaissance and intelligence-gathering purposes. These early satellites and a plethora of others that followed mapped and studied the geography, geology, and atmosphere of Earth.

More recently, reduced costs and highly improved image resolution have made these imaging satellites as much desired as communications satellites because of their dual-use applications. The same satellite that takes topographical images for agricultural uses can just as easily keep track of another country's military and/or industries. Today, remote-sensing satellites use either visible light (i.e., a photograph) or non-visible light methods, such as x-ray or infrared. Such satellite imagery has many practical applications for national economic development, such as in land accretion studies, land use mapping, forest inventory, surface water distribution, river course monitoring, crop identification, and general cartography. In addition, real-time meteorological satellite data are used for population security in severe weather prediction, storm surge estimation, and disease identification. A multitude of satellites also document the effects of pollution and global climate change, both for long-term predictions and for short-term effects on population centers and agricultural production. For these and a host of other reasons, interest in acquiring satellites and their attendant benefits has been growing steadily among developing countries (see Table 3.1).

To further these interests, the growth of space programs among the non-traditional space actors has been impressive and expands by the year. As of 2010, there were no less than 20 countries that had developed

Table 3.1 Satellites launched by developing countries

ESA – European Space Agency

Country	Year first satellite launched	Launch origin	Manufacture origin	Purpose of first satellite	Total satellites launched (as of September 2011)
China	1970	self	self	demonstration	108
India	1975	Russia	self	science	46
Indonesia	1976	USA	USA	communications	11
Brazil	1985	ESA	Canada	communications	12
Mexico	1985	USA	USA	communications	7
Argentina	1990	ESA	self	communications	8
Pakistan	1990	China	self	communications	2
Thailand	1993	USA	USA	communications	7
Turkey	1994	ESA	France	remote-sensing	5
Ukraine	1995	self	self	remote-sensing	1
Chile	1995	Russia	USA	remote sensing	1
Malaysia	1996	ESA	USA	communications	6
Egypt	1998	ESA	EU	communications	3
South Africa	1999	USA	self	communications/science	2
Morocco	2001	Russia	self/Germany	communications	1
Algeria	2002	Russia	self/UK	Earth observation	1
Iran	2005	Russia	self	Earth observation	1
Kazakhstan	2006	Russia	Russia/France	communications	1
Colombia	2007	Russia	self	communications	1
Vietnam	2008	ESA	USA	communications	1
Venezuela	2008	China	China	communications	1
Bangladesh	2013 (announced)	to be determined	to be determined	communications	–

* Source: SATCAT Boxscore, www.celestrak.com/satcat/boxscore.asp

enough launch capacity and sophistication to put payloads into orbit, mostly for research purposes but also with missile and satellite development applications. Three-quarters of these states are classified as developing countries according to the principal criterion used by the World Bank, which sets the threshold at a GNI per capita below US$11,905.[8] Three developing countries—Brazil, China, and India—have committed themselves heavily toward the development and utilization of launchers for orbital and even geosynchronous orbital satellite placement, and at least nine other developing states have announced plans to join them in developing independent space launch capability within the 2011–15 time frame.

One indicator of the growing space presence of EMSAs is the number of communication satellites owned and manufactured or co-manufactured by them. For example, of the 632 communications satellites in orbit in 2009, 70 (11 percent of the total) are owned by national governments, private companies, or government-private organizations based in EMSAs. More than 300 more communications satellites are projected to be launched globally by 2016, of which an increasing percentage will come from EMSAs. Similarly, of the 17 satellites launched in 2007 for designated scientific purposes, six were owned by EMSAs.[9] Concurrently, 27 countries have remote-sensing capabilities compared with a mere three countries 30 years ago; 25 of these countries have official space programs. Space policy has become integral, if not a *sine qua non*, to capable states' national security and developmental policies, and collectively, developing countries now comprise the greatest area of growth in space activities.

Classification of emerging space programs

The space programs of developing countries can be divided into three distinct groups, whose membership is based upon the capacity to develop and carry out space-related programs. This categorization is derived from a more basic scheme developed originally by Newberry (2003), which he used to examine whether Latin American states' forays into space-related endeavors were beneficial or detrimental to US security interests.[10] In order to better understand the genesis, evolution, and future trajectory of the space programs of EMSAs, it is useful to explore them through the prism of space power that was explored in Chapter 1, because space activities today address so many of the crucial areas of importance to the modern nation-state.

The most advanced space actors in the developing world are 'first tier' states, which have achieved the capability to autonomously produce space technology, have developed (or are on the cusp of developing) indigenous launch capability for both orbital and geosynchronous satellite placements, and have national space agencies, and whose space programs evolved from research and development (or attempted development) of ballistic missile and nuclear programs. The 'second tier' states are those that produce some of their own space technology, have basic launch capacity (typically sounding rockets), have national space agencies, and frequently, out of necessity, collaborate with more advanced states' programs in the production of space technology. The 'third tier' states occasionally make contributions in space-related technology, almost always purchase space-related technology from more advanced producers, and almost always collaborate with other more developed space actors to achieve their space policy goals. Rather than being space-faring, third tier space actors have made the policy decision to invest in space technology to accomplish what could not be done otherwise. Applying these criteria more broadly across the entire spectrum of developing countries, we can categorize all developing states that have initiated space policies, and by extension, more fully understand how their space policies fit into their broader foreign policy goals.

First tier states

Three developing countries meet the general criteria of a first tier EMSA: Brazil, China, and India. It is no coincidence that these three aspiring space powers are also three of the four fastest-growing developing economies of the post-Cold War period. Together with Russia, they comprise the so-called BRIC group. These three EMSAs are the most likely candidates to join and/or compete with the established space actors in the near future because of their size, growing economic footprint, and political influence, and the complex histories and policies of their missile and nuclear programs. For these reasons, these most successful of the EMSAs merit in-depth analysis.

For these largest and most advanced EMSAs, their space programs contribute to comprehensive national power (CNP), an idea attributed in part to classical Chinese political thought. CNP takes into account a broad spectrum of sources of power, more than the sum of mere hard and soft power, and incorporates many of the stratagems outlined in Sun Tzu's timeless *Art of War*, such as image, prestige, and deception.[11] The creation of space programs by EMSAs contributes to CNP by creating and promoting a broad array of capabilities in space-related technologies and abilities. Among these current, and perhaps a couple of other potential first tier EMSAs, a number of other commonalities exist. In each area, one can see evidence of the trails blazed by the older, more established space actors. Besides having relatively higher budgetary allocation for space-related activities than other EMSAs, first tier countries also share the following characteristics: (1) a history of the pursuit of ballistic missile production; (2) the development of a nuclear energy sector and an indigenous armaments industry, and the pursuit of a nuclear weapons program (fully realized or not); (3) a national space agency; (4) the construction of national launch facilities along with the domestic development of both satellites and launch vehicles; (5) the endogenous development of satellite technology and associated resources that complement each state's national development and security needs.

Three characteristics are notable in the development of space activity in these states. First, the story of each first tier space program, like that of practically all major space powers, traces its lineage back to the policies, ambitions, and development of ballistic missile systems. Reflecting the strong synergy that has always existed between ballistic missiles and space launch vehicles, the militaries of first tier countries have invested substantially in missile development and have all achieved sufficient success to become exporters of missiles systems and their associated technologies. This aspiration and technological advancement have led directly to the acquired independent launch capability of first tier space actors.

Second, first tier countries have not only possessed nuclear ambitions but have also achieved the partial or full development of programs in nuclear weapons and/or nuclear energy. This follows the previously established models of nuclear proliferation which demonstrate that countries capable of producing such weapons will do so because of the implied security as well as the internal bureaucratic dynamic that emerges from larger, wealthier states seeking status and further socio-economic gain.[12] In two of the three cases examined herein—China and India—the need to extend the reach of the nuclear weapons program was the understood, though undeclared, motivation for space program development. In the third case, Brazil, a nuclear weapons program was initially undertaken and developed to the cusp of implementation, subsequently abandoned after military rule, and then quietly restarted in the first decade of the twenty-first century in the form of a nuclear submarine program and a growing nuclear power industry. Concurrently, each first tier EMSA possesses an extensive domestic armaments industry that has, at various times over the past half century, both provided weapons to their national militaries and also made each country a net weapons exporter.

Third, while a multitude of countries have established an official governmental agency or office to coordinate space-related activities, the first tier states have all established state space agencies that plan and coordinate national space activities, in addition to possessing domestic space industries, indigenous launch programs, and a home-grown defense industry that has contributed to the technology base and the strategic rationale for creating space programs. In addition to their national space agencies, each first tier country also benefits from a number of associated scientific agencies, commercial aerospace companies, and research organizations largely or solely committed to researching and developing space-related technologies.

China's long march into space

During the first decades of the space era, space programs were considered the special province of the two Cold War superpowers and a select few of their closest allies. But the post-Cold War period, and especially the twenty-first century, is strongly trending toward becoming a more diversified era in space, as developing states, particularly those in Asia, join the space club through a variety of ambitious space goals. Without a doubt, the leader among these emerging space actors is China, which in 2003 became the third country to independently put one of its citizens into space.

China began its space program more than a decade after the Soviet Union and the United States. But while a relative latecomer to space, China has nonetheless laid out ambitious plans that mirror its rise as an economic power as well as the evolutionary trajectory of the space actors that preceded it. China's approach to its space program has been highly politically charged and motivated, with close links to national security as well as socioeconomic development goals. Initially, China space efforts were an archetypal example of a security-oriented program in which the military tail wagged the space dog, but in recent decades socioeconomic priorities have been elevated as well. Thus, China's growth as a space power is meant to support the priority that Beijing places on rapid and sustained economic growth through ever-improved technology in sciences and industry.

China's space program is distinguished by having a full range of capabilities typically found only in DVSAs. These capabilities include satellite design, commercial and military launch services, and most recently, human space flight and deep space robotic exploration. China's space program serves both its national security and human and economic development interests, which are meant to ensure the country's segue into a position of international leadership in the twenty-first century. In addition to more traditional security concerns, China's space policy has also emphasized the socioeconomic benefits of space applications in areas as diverse as meteorology, environmental protection, disaster monitoring, water conservation, and topographical mapping, all of which serve the national economy and national security equally well. Indisputably, China is the emerging space actor closest to shedding its "emerging" classification.

The road toward the creation of the Chinese space program has been built upon the country's 2,000-year history of rocket experimentation. Not only was gunpowder invented in China but a dizzying array of rockets was conceived for practically every possible application. For example, over 1,000 years ago gunpowder rockets called "fire arrows" were an integral component in Chinese military arsenals. In 1232, the Chinese used a barrage of "arrows of flying fire" against invading Mongols, marking the first documented use in war of solid-propellant rockets. By late 1500s, Chinese military forces were already using an early, albeit crude, multi-stage rocket. Called *Fire Dragon over Water*, the rocket was an unso-phisticated precursor of the modern cruise missile, and its multi-stage design presaged exactly what would appear four centuries later to achieve escape velocity.[13]

What are the specific motivations for China's ambitious leap into space? Part of the explanation can be derived from the necessity of servicing the economic needs of the country's 1.3 billion people and from China's ascension as a world economic leader. These two factors alone suggest that finding and controlling resources, today done with the help of satellite-based topographic mapping, is understandably a highly important objec-tive for the Chinese space program. Another important incentive behind China's rush into space is that there are a limited number of orbital slots for satellites in geosynchronous orbit (GEO) and a state's access to these slots can only be assured by occupying the slot in GEO with a satellite. In an era of satellite communications, reconnaissance, and information systems, for a state not to have claimed an orbital slot is tantamount to abdicating its national power, perhaps even its sovereignty in the longer term.

Simply put, China's space program serves the foreign and domestic policy goals of the Chinese state. Equally important is the fact that China's recent economic and political expansion is perceived by Chinese strate-gists as possibly being checked by US hegemony in the Pacific region. Beijing considers the US to be an obstacle to China's aspirations of being the unchallenged regional hegemon in Asia. China's military planners have plans to create space-based surveillance and the other dual-use space technology to establish regional parity with the United States within the Asian sphere.[14]

As in the US and the Soviet Union before, the road to China's space program was paved by developments in the country's missile and nuclear programs, which themselves were the outgrowth of the country's geopoliti-cal situation and history. Having endured the seizure of Hong Kong by the British, the Japanese invasion and brutalization of Manchuria, the upheaval of the 1949 Chinese Revolution, and finally, the near-encirclement of its territory by perceived enemies in the form of US mili-tary bases in East Asia, a sometimes antagonistic Soviet Union to the north, and a wary and uncooperative India to the south, China's leaders pursued what they considered the best course of action to ensure their country's sovereignty.

The development of China's missile and space programs can be divided up into three distinct phases, each one indicative of a policy shift reflective of the geopolitics of the time. The first phase was a period of learning, in which China began to recognize the utility of missile and space technology in the service of national interests. The birth of these programs also paral-lels the US experience, in that both countries had similar reactions to *Sputnik*. As impressed and disquieted as the American policymakers were, the Chinese reaction was analogous and built upon an already decided path toward space.

Having greatly appreciated the prestige effect of *Sputnik* and the rest of the Soviet Union's early nuclear and space achievements, China yearned for membership in these two clubs, which Mao believed would endow the country with great-power status. It was thought that nuclear weapons and ballistic missiles would be a clear sign of China's achievement of military strength, which could boost the Chinese people's self-esteem and translate into international influence and prestige.[15] With Soviet assistance in nuclear technology and an already antiquated Soviet R-1 missile (a copy of the German V-2) provided by the USSR, China embarked on its rocketry modernization program. In many cases, the Soviet-provided rocket donations were "reverse-engineered" (i.e., disassembled, studied, and then locally produced) both to generate indigenous copies and to work towards production independence.

Undoubtedly to the chagrin of modern American policymakers, in an important and ironic way the United States unwittingly helped to further China's ballistic missile and space programs by committing what many historians now consider one of the greatest strategic blunders of the twentieth century. One vignette is worth telling because it reveals some of these characteristics as well as the substance of China's drive to become a respected world power, which would necessarily include becoming a space power.

At the height of McCarthy's "red scare" witch-hunts in the United States, Chinese-born but US-educated Qian Xuesen (sometime transliterated in the US as Tsien Hsue-shen), the Robert H. Goddard Professor of Jet Propulsion at the California Institute of Technology with the reputation as one of the leading rocket scientists in the United States, was accused in 1951 of harboring communist sympathies. Qian was a protégé of Theodore von Kármán; his many achievements included research and innovative designs that would eventually influence the design of the US space shuttle. But despite having been one of the founders of the US Army's Jet Propulsion Laboratory (JPL) during World War II, having been integrally involved in the debriefing of Nazi scientists in Operation Paperclip (even interviewing Wernher von Braun), and having received the civilian-equivalent rank of colonel in the US Army, Qian was put under near house-arrest without trial for five years. After much diplomatic wrangling between the US and China, Qian was finally allowed to leave. Having been discarded by his adopted country, and likely (and understandably) disgruntled, Qian accepted a deal in 1955 to be exchanged for US prisoners of war held by China from the Korean War. He returned to China, where he declared his intention to help China modernize.[16]

Qian assumed the role of chief architect of the Chinese missile program, where his work was directly responsible for the creation of China's ICBM ballistic missile program, its famed *Silkworm* cruise missile, and its still-used Long March launch vehicle family.[17] To promote these programs, the then-classified Fifth Research Academy of the Ministry of National Defense was established (following the Soviet tradition of giving national security projects nondescript names to confound outsiders). The Fifth Research Academy was assigned over 160 of the country's top university graduates in science and engineering, many of whom became Qian's protégés. With the brainpower in place, attention turned to building the technology.

Much like France's post-World War II decision to pursue nuclear weapons and launch platforms to ensure French territorial integrity, in 1955 the Chinese Communist Party (CCP) declared its intention to build a nuclear device under the code name 02. The next year the Missile and Rocket Research Institution was founded, with agreements with the Soviet Union for technical support and technology transfers in the nuclear and missile fields. With its national defense priorities now institutionalized, and at the suggestion of its Soviet advisors, China began its indigenous ballistic missile program in the mid-1950s.

Appropriate to its defense-related inspiration, the early Chinese space program was originally placed under the People's Liberation Army (PLA). This should not be a surprise given the shaky security environment of the time. Largely isolated by the international community, facing potential threats from the United States, and hampered by technical restrictions imposed by the West, China was understandably inclined to delegate the development of rocketry and space-related technologies to the PLA. The Soviet Union played a key role in the initial stages of China's space development through technology transfers. Using Soviet plans and 14 Soviet R-2 rockets (improved versions of the V-2), which had been provided under terms of the 1957 Sino-Soviet New Defense Technical Accord, China's ballistic missile program was off the ground.[18] China's first task was to copy the R-2 under the code name Project 1059, which was an effort that benefited from close Soviet assistance but, at the same time, suffered from China's lack of expertise and infrastructure to provide even the most basic materials.[19]

To accompany the forthcoming Chinese rocket, Mao insisted that "We must make artificial satellites, too."[20] Chinese leaders were thoroughly impressed by *Sputnik*, but China had already made great strides toward implementing an active space development policy even before the Soviet success. Nonetheless, Mao had also recognized the potential dual-use nature of Soviet heavy-lift rocketry. So at the Eighth Congress of the CCP in May 1958, Mao announced China's intention to build the first Chinese satellite.[21] He purportedly declared that "if we're going to throw one up there, then throw a big one … something like that chicken egg of the Americans won't do!" Based on this edict, the Chinese Academy of Science prioritized the satellite program and created Group 581, which was charged with a three-part plan: 1) build a sounding rocket; 2) launch a

200-kilogram satellite; and 3) launch a satellite weighting several thousand kilograms.[22]

Mao's plan not only put satellites on the national agenda, but also outlined a roadmap for the future development of ICBMs, nuclear submarines, nuclear weapons, and even more advanced rocketry, all of which were to be completed as soon as possible. However, domestic and geopolitical events would delay that vision from becoming reality. China's quest to join the embryonic space club by launching a satellite was constrained by many economic and technical limitations, not to mentioned internal and international influences that came to mold China's space program. One great impediment was that Mao's space pronouncement coincided with the beginning of the Great Leap Forward program, the massive, and ultimately disastrous, experiment in economic and social engineering that was meant to make China an advanced industrial state within 15 years.

The previously helpful Soviet partner became less so after *Sputnik*, in part because the USSR came to interpret China's space ambitions as potential competition for the leadership of communist influence in the world, and in part due to Mao's calculated insults toward Soviet premier Nikita Khrushchev.[23] In due course, the gradually worsening relations caused by the Sino-Soviet split of the 1960s meant that China would have to assume full ownership of its rocket program. This signaled the second phase of the Chinese space program.

In the midst of its technological and ideological isolation from the USSR and the US, China dove headlong into creating self-reliance in missile technology. Despite numerous bureaucratic, economic, and international obstacles, progress was made, albeit slowly at first. Using a bicycle pump to pressurize the fuel tank, the first Chinese liquid-fueled rocket, the T-7M, was successfully launched in September 1960.[24] But from that first tentative step, progress was almost geometric in its rapidity. Within five years, China had both successfully tested its first nuclear weapon (October 1964) and tested the DF-2A (November 1965), a medium-range ballistic missile able to carry a nuclear warhead. These two innovations signaled the beginning of China's modern integrated program (best-guess US intelligence estimates are that China currently has about 240 nuclear warheads divided between a modern triad of land, air, and sea forces[25]).

However, countering these advances was the internal political turmoil of the Cultural Revolution (1966–76), which decimated China's intellectual and scientific communities as Mao attempted to wrench the last vestiges of Western influence from the country. Fortunately for China's space program, much of the space scientific community emerged relatively unscathed. A lasting legacy of the Cultural Revolution was a number of significant changes in the division of power and labor in China's space organization. Perhaps the most important was the division between Beijing and Shanghai, in which the two cities vied for production dominance in launch vehicles and satellites.[26] This wary relationship continues to this day and speaks to the predominance of political, rather than scientific, concerns within the Chinese space program.

Materializing from a grueling 42-day meeting in 1965 was the institutional framework for the Chinese space program, a structure that remains largely intact to this day.[27] The Academy's primary goal, called Project 651, was to launch China's first satellite within two years. Then, inspired by the imminent US moon landing and its attendant prestige boon, Mao committed China to a formal space program in 1968, and the Chinese Academy for Space Technology was founded.

Decided in 1971, Mao's ambitious plans for space included a manned space program, the top secret Project 714, which aspired to put Chinese astronauts (called *yuhangyuan* in Mandarin or sometimes "taikonauts" in the West) in space by 1973 aboard the *Shuguang 1*, a copy of the Soviet *Soyuz* capsule. The program even progressed to the point of choosing 20 potential astronauts from the air force, who trained with cardboard and wooden spacecraft mockups. The program was cancelled in 1972 because of financial constraints as well as the Cultural Revolution, with Mao proclaiming that "we should take care of affairs here on Earth first, and deal with extraterrestrial matters a little later."[28] Despite this setback to the manned program, the unmanned space program continued unabated.

In April 1970, China became the fifth country to successfully put its own satellite into orbit, though it contained German electronics and benefited from French technical assistance.[29] The *Dongfanghong-1* was set aloft by a *Long March* 1 rocket, itself a modified CSS-3 intercontinental ballistic missile with a modified third stage, capable of placing up to 300 kilograms in low Earth orbit (LEO). From the beginning, the satellite addressed the prestige-building agenda of the Chinese space program, sending out the patriotic song "The East is Red." The satellite's official slogan of "get it up, follow it around, make it seen, and make it heard" emphasized its public relations rationale. A shiny metal ring was even installed on the satellite to make it easier for Earth-bound observers to find it as it crossed the night sky. The Long March series of launchers have since been used 133 times (as of April 2011), and with a success rate of 94 percent, China has become one of the world's most dependable provider of satellite launch services.

Following this first successful satellite launch, Chinese leader Zhou Enlai outlined China's foreign policy priorities, all of which would benefit from China's enhanced missile and space programs: (1) the reunification of Taiwan with the mainland; (2) the elimination of US military bases from Asia; (3) the withdrawal of the sizable Soviet military forces positioned along its common border; and (4) the deterrence of Japan's reemergence as a military power.[30] In 1975, the development of communications and reconnaissance satellites was included as an objective in the State Plan.[31] That year, China became the third country to independently launch and recover a satellite.

The third and current period of the evolution of China's space program began with the Third Plenary Session of the CCP Central Committee in 1978. With the country's economy on the brink of collapse, the decision was made to prioritize economic and social development for the foreseeable future. Integral elements of this policy switch were to shift some military production from the defense industry toward a civilian-based production model and to re-engage the international community, thus eschewing the country's traditional isolationism and embracing radical new approaches to both its international relations and the direction of the space program. During the leadership of Deng Xiaoping (1978–92), the entire defense sector, which included the space program, was scaled back and instructed to develop technology with commercial applications.[32] Even engagement with the United States became a necessary evil. From 1978 to 1980, Chinese delegations traveled to the US to try to buy a communications satellite from its former foe; the deal fell through, though not for want of trying.[33]

This latter engagement in internationalism has become a hallmark of modern Chinese space policy. During Deng Xiaoping's rule, a new, highly pragmatic program called the "Four Modernizations" was conceived. It used an innovative calculus for estimating China's strengths based on agriculture, industry, technology, and defense. It also emphasized comparative restraint in the use of force while simultaneously increasing efforts to modernize China's military.[34] First proposed in 1964, this version of the modernization agenda was designed to make China a world power by the early twenty-first century. However, since the country's space program at the time provided little economic value, it was, for a time, left in the dust of the new drive for economic growth. Some factories that had produced space-related technologies were converted into consumer and industrial production centers.

But despite the obstacles, the 1980s saw some of China's boldest moves toward becoming a modern space power. In 1983 China became a signatory of the Outer Space Treaty, and throughout the 1980s China pursued adhesion to a number of other space-related multilateral agreements. China also began to aggressively seek out partnerships in a wide range of projects with more developed space actors as well as some less developed space actors. These projects included satellite production and launch services, the construction of tracking facilities, data processing, and even providing space technology training for experts from less developed countries.

Like the birth of its overall national space program, China's growth in the commercial satellite business was unintentionally provoked by US foreign policy. After the Chinese government's fierce crackdown in 1989 at Tiananmen Square, President George H.W. Bush imposed economic sanctions on China, which included a prohibition on the export of dual-use satellite technology. Additionally, the US Congress passed the

Tiananmen Square sanctions law, which suspended the export of US satellites for launch by China. After a decade of domestic political bickering, the technology export controls law was modified again in 1999, this time transferring licensing authority to the US State Department, which resulted in a much more stringent export control regime on dual-use satellite technology. Satellite companies in the US were hamstrung in their ability to compete for major international projects, but China turned a potential roadblock into an opportunity.

Before 1999, the US was the overwhelming leader in the commercial satellite-manufacturing field, with an average market share of 83 percent. After that time, market US share declined to 50 percent (with an estimated loss of US$1.5 to US$3.0 billion to the US economy).[35] While this cannot be blamed entirely on changes in export regulations, they nonetheless contributed to the decline. Since the change in US export policy, no Chinese satellite operator has chosen to purchase any satellite that is subject to US export regulations; instead they have selected European and Israeli suppliers, with six different satellite orders since 1999. Additionally, China has made a commitment to the building of its commercial satellite bus, the DFH-4, by the China Academy of Space Technology.[36] This bus has been successfully marketed to other countries that feared the possible complications resulting from US export policies, including Nigeria and Venezuela.

To address the need for greater nimbleness and flexibility, the formerly centralized space industry was semi-privatized. Responsibilities for launchers, satellites, and space technology were distributed among a number of semi-autonomous, state-run enterprises. Research and development among these companies, which before 2000 accounted for less than one percent of GDP, are on track to rise to 2.5 percent in 2020.[37] Likewise, Chinese scientific publications are on the rise from just 2 percent of the world total in 1995 to almost 15 percent in 2010, thus becoming the second largest contribution in the world behind the United States (at 22 percent).[38] Similarly, the formerly reclusive communist state publicly announced in 1982 at a space conference in Switzerland its intention to enter the commercial satellite market.[39] Principal among the new corporate entities that would usher in the new space age was the Great Wall Industry Corporation (GWIC), which was founded in 1980 and authorized by the Chinese government to promote and provide its launch services to the world satellite market.

The GWIC is also in charge of satellite technology trade, acting as the principal marketing channel for China's mushrooming aerospace industry. Since the Long March launch vehicle family first came into service in 1970, the GWIC has established business relationships with more than 100 companies and organizations throughout the world.[40] The creation of the GWIC additionally served to end the traditional management of industry by the PLA, which for decades had managed the most critical manufacturing firms and produced upwards of US$3 billion annually for itself. The Divestiture Act of 1998 wrested control of these purely commercial activities from the PLA, which was intended ostensibly to restore the PLA's professional discipline but in fact was intended to reassert government dominance over what many considered to be an institution rife with corruption and ill-prepared for the demands of the professional military of a hegemonic state.[41]

With the PLA extracted from its erstwhile position as soldiers-in-business (*bingshang*), the more rapid commercialization of the space industry ensued. Nonetheless, the state maintained an important direct role through state-associated enterprises that emerged, such as the China Aerospace Science and Technology Corporation and the Chinese Academy of Space Technology: each is a semi-autonomous but nonetheless integral cog in the Chinese policy to use space for national development and security enhancement. In addition, the civilianization of the space program did not necessarily mean the end of China's use of its space assets to further the country's terrestrial agenda of supporting the ambitions of so-called rogue states. Since the early 1990s, the GWIC has been implicated in the proliferation of Chinese missile technology to Iran, North Korea, and, until 2003, Iraq.[42]

China successfully put its first indigenously manufactured communications satellite, *Dongfanghong-2* ("The East is Red 2"), into geosynchronous orbit in 1984. On its heels, two years later, the State High-Tech Development Plan (SHTDP; also known as Plan 863) was conceived as a government-led force to stimulate massive improvements in biotechnology, information technology, lasers, automation, energy, and aerospace that would propel China into parity with other technologically advanced states.[43] Aerospace emerged as a primary development area, and world events helped to promote China's emergence as a rising world space power—the 1986 *Challenger* space shuttle explosion, together with a string of failed launches of ESA rockets, made Chinese launch services more attractive to businesses.[44] Economic rationality became the prevalent mentality, and members of China's space community and government were ever more entrepreneurial.

Implicit in the SHTDP was the restarting of China's manned space program, based on the premise that China's ability to lay claim to space power would be predicated on displaying national strength in technology, which would yield positive international regard. At the same time, the resounding success of space-based assets utilized by the US military during the 1990 Gulf War for communications, munitions guidance, navigation, and reconnaissance underscored for the Chinese leadership the need to overhaul the country's mass-army organization, which was antiquated in the modern, technologically driven security environment. The need for space to serve the national security needs of the Chinese state became more acute than ever.

In 1986, the SHTDP was implemented, in part to free China from external technology dependencies.[45] This same year, China took advantage of the aforementioned troubles in the European and American space programs and began to offer commercial launch services. Beginning with satellite launches for Hong Kong (pre-reunification), Saudi Arabia, Sweden, and Australia, China aggressively sought to become the launcher of choice for companies and governments from Asia, Africa, and South America. In 1988, China launched its first of two meteorological satellites (the second in 1990), becoming the third country in the world to produce and put into polar orbit an indigenously developed meteorological satellite.[46] Following these successes, the Chinese government decided to pursue an even more expansive space program, which would coincide with the country's export sector shifting from lower-return primary and labor-intensive exports (e.g., textiles) toward more high-tech, higher-profit exports. Such juxtaposition of developmental and space policies coincided with the country's growing revenues available for capital-intensive space projects.[47] China could now afford to become a player in space, and that is exactly what it did, entering into the international commercial satellite launch market after the mid-1980s. After a series of accidents, China imposed upon itself international launch standards to bring its program in line with foreign commercial expectations.

China's emergence as a competitive launch provider as well as a budding space power had obvious effects on the United States, which tried briefly to impede China's expansion into commercial launches by attempting to restrict the technologies that US companies could sell China.[48] In addition, under the umbrella of non-proliferation, the US government restricted American companies' commercial payloads from being launched by foreign providers. This specifically targeted China and Russia for the public reason of concern about technology transfers as well as the Tiananmen Square massacre; however, the real reason for the policy had more to do with US launch providers' disgruntlement over a loss of business to these two countries.[49]

But despite these external attempts to frustrate its space program, China has become the third largest national provider of launch services, after the United States and the European Space Agency (ESA). From 1987 to 2010, China performed over 30 international commercial satellite launches, with an average price of US$40 million per launch (about 60 percent lower than the average cost of a launch by the ESA).[50] China's domestic space industry is well developed and diversified. The China Aerospace Science and Technology Corporation is a state-owned company that builds a variety of communications, weather, science, remote-sensing, and navigation satellites. The second largest contractor is the Chinese Academy of Space Technology, which focuses on commercial broadcast and information satellites.

Another important aspect of the growth of China's space program is its expanded cooperation with space actors other than the United States. Such "space diplomacy" is a calculated approach to draw other space actors closer though science and technological diplomacy (*keiji waijiao*) and to reduce China's dependence on US technology and its associated technology restrictions.[51] Beginning with the launch of its first satellite, China has worked extensively with the European Space Agency. Since 2000 this cooperation has blossomed to include joint scientific missions to study Earth's magnetic field (2001) and develop better remote-sensing technology (2004), a Sino-German solar telescope, and the purchase of French-built satellites to complement the large home-grown satellite industry. China's cooperation with some Latin American countries has paralleled its economic investment prowess in the region, which reached US$23 billion in 2010.[52] In this region, China also gave substantial support to the space aspirations of one of the United States' most vocal critics, Venezuelan president Hugo Chávez, with an agreement to jointly build and launch the South American oil giant's first telecommunication satellite, *Venesat-1* (launched 29 October 2008). In Africa, China has been very generous in bankrolling—with Nigeria's petroleum as collateral—the lion's share of Nigeria's costs for its two communications satellites *NIGCOMSAT-2* and *NIGCOMSAT-3*. In Asia, China has been at the forefront of the eight-state Asia-Pacific Space Cooperation Organization, which seeks to build reconnaissance satellites for disaster monitoring, and has been instrumental in expanding the Association of South East Asian Nations (ASEAN) from merely an economic forum to one that actively includes space-based technology in its agenda. By 2007, China had signed 16 space-related agreements with 13 different governments and organizations, and had established space cooperation relationships with more than 40 countries and international bodies.[53]

The biggest leap forward for China's space program is the resurrection of its manned program. As previously described, manned flight had been first proposed and tentative steps taken as far back as 1966. However, it was not until February 1987 that the manned option would officially resurface in response to the US conventional weapons buildup during the Reagan administration. In 1988, a group of 17 Chinese space experts met and debated the future of China's space program. The possibility of a Chinese space shuttle was proposed by Institute 601 of the Air Ministry, but was eventually shelved in favor of a simpler manned program. Only a month after the Tiananmen Square massacre, a July 1989 "expert commission" report advocated building a manned capsule with a maiden flight date of 2000; it also recommended the simultaneous development of a reusable space shuttle, though this proposal was once again shelved.[54] The Air Ministry set up a manned space program office, and through a maze of internal opposition and Russian foot-dragging on proposals to buy a *Soyuz* capsule, a three-step plan finally emerged which established the path to manned flight.

Originally carrying the non-descript title Project 921, and later renamed *Shenzhou* ("Sacred Vessel" in Mandarin), the Chinese manned spacecraft was modeled after the Russian *Soyuz* but was about 13 percent larger, possessed four separate engines (compared to *Soyuz*'s two), and enjoyed better electronics than the Soviet vehicle.[55] China declared that it planned to use the *Shenzhou* to develop orbiting skills and technology in a project consisting of three progressive steps. The first was to be five unmanned test flights of the *Shenzhou* spacecraft. For maximum propaganda benefit, the first flight in 1999 (*Shenzhou 1*) coincided with the fiftieth anniversary of the PRC as well as the return of the Portuguese colony of Macao to China, and it carried a mannequin and a host of nationalist paraphernalia. The second flight in 2001, *Shenzhou 2*, carried a monkey, a dog, a rabbit, and some snails, which stayed in orbit for seven days before parachuting back to ground in Inner Mongolia. The third and fourth Shenzhou flights in 2002 carried mannequins.

Finally, on 15 October 2003, China's first manned flight took place aboard *Shenzhou 5*. Though the astronaut, Yang Liwei, spent less than one day (21 hours) in space, completing 14 orbits, the feat put China on the map as only the third country to independently put humans in space. Two more manned flights followed in 2005 and 2008—*Shenzhou 6* and *Shenzhou 7*—with two- and three-person crews respectively. The *Shenzhou 7* mission also featured China's first spacewalk, as well as the accompanying launch of the *BanXing* (BX-1) companion satellite, which sent back images of the spacewalk. The mission was a resounding, though calculated, success. It undoubtedly furthered the Chinese space program and proved its continued advancement to the world, and, like other programs before it, helped to foster the development of even more advanced technologies. But the mission's timing and its broad-reaching political goals had equal effects.

Occurring between the Beijing Olympics and the Chinese National Day, after the disastrous Sichuan earthquake, and during government protests and persistent high inflation, the image of a spacewalking *yuhangyuan* provided an inspirational moment for an increasingly restive population in a non-democratic system still experiencing popular discontent.[56] The images of Col. Zhai Zhigang and Lt. Col. Liu Boming floating in space, tethered to their capsule, undoubtedly helped to assuage temporarily some of the disgruntled Chinese population and lent further credibility to the CCP's absolute domination over the system, as well as invoking a strong sense of national pride and international status for the country. Besides being a very high profile patriotic event—analogous in China to the nationalistic fervor produced by Yuri Gagarin in the USSR—it served as a demonstration of China's growing technological virtuosity, which it was hoped would provide a boost to Chinese high-tech exports and give the country greater diplomatic as well as technological parity with Japan and the West.[57] Moreover, like *Sputnik* and other space firsts, this mission and China's other manned missions served as a bold notice of the country's coming of age and its commercial launch capability.

Shenzhou 7 was, in sum, an economic, domestic, international, political, and strategic project all rolled into one, though it also created a minor stir among scientists and policymakers. During its orbit, Shenzhou 7 and the BX-1 passed uncomfortably close (45 kilometers) to the International Space Station. This led some observers to speculate that China was practicing co-orbital anti-satellite interception procedures, and to cite this event as more evidence of the Chinese government's dual-use space policy and its geostrategic gamesmanship, what Joan Johnson-Freese (2004) called China's "Space Wei Qi" (in reference to a Chinese strategic game similar to chess). Others, on the other hand, have argued that the West is inferring Chinese motives that simply are not there.[58] Perhaps to understand Chinese motivations, one needs look no further than the Chinese proverb: "When riding a tiger, it is difficult to get off," which is to say that China pursues its continued space development because it feels it has no choice given the realist history of space politics.

The second step in the manned program will entail a series of flights to conduct rendezvous and docking operations in orbit for a smaller eight-metric-ton Chinese space laboratory, *Tiangong I*, which will form the core of a 20-metric-ton Chinese space station, scheduled to be put in orbit in 2020.[59] The space lab will be lifted into orbit aboard the new *Changzheng-5* heavy-lift launcher, which will be the world's second most capable launcher (after the US *Delta IV Heavy*). Called Project 921–2, the mission has led to the Chinese government extending invitations to Canada, the ESA, and Russia to participate in cooperative efforts revolving around the space station. The United States is notably absent from the list of invitees, presumably because of the earlier US rejection of a Chinese request for a place on the International Space Station.[60] The first phase was conducted with an unmanned space module, launched into orbit on 29 September 2011, which was meant to capitalize on the termination of the US space shuttle program.[61] The program will ultimately include a more advanced space laboratory, a cargo ship (*Tiangong 2*), and a number of smaller research modules (*Tiangong 3*).[62]

By October 2000, China had launched over 100 of its own satellites, with a flight success rate of over 90 percent. The satellites were of four different types: (1) meteorological; (2) scientific research; (3) telecommunications; (4) recoverable military reconnaissance (see Table 3.2). Current plans call for almost 60 more government launches during the 2010–20 time frame.[63] In addition to the large military and prestige components of China's space program, there are tangible socioeconomic problems and goals that the Chinese government seeks to address through the space program, particularly in environmental protection, reducing the widening wealth disparity, and promoting renewable energy sources.

Table 3.2 Chinese satellite series

Satellite series	Purpose	Orbit	Additional information
Feng Yun	meteorological	Polar and GEO	infrared, radar, and visible
Beidou	navigation	GEO and MEO	real-time passive 3-D geospatial positioning; eventually to include 35 satellites
Yaogan Weixing	land survey, agriculture, and disaster monitoring	LEO	all-weather imaging with military applications
Tianlian	data relay	GEO	space launch support; manned mission support
Shentong	communications	GEO	military and government communications
Fenghuo	military communications	GEO	digital
Ziyuan	remote sensing	LEO	jointly developed with Brazil
Dongfanghong	telecommunications	GEO	voice, data, TV
Shijan	science	LEO	scientific experiments
Haiyang	oceanography	GEO	ocean observation

Source: Daphne Burleson, *Space Programs Outside the United States: All Exploration and Research Efforts, Country by Country*, Jefferson, NC: McFarland & Company, 2005, available www.sinodefence.com/strategic/spacecraft/default.asp and www.globalsecurity.org/space/world/china/index.html

Note
GEO – geosynchronous orbit; LEO – low Earth orbit; MEO – medium Earth orbit.

The eleventh Five-Year Plan of 2006–11 specifically mentions the need for the country to improve medical care and education in rural areas which will be provided, in part, through satellite-based telemedicine and education. Another example of the socioeconomic application of the Chinese space program is the development of the Feng Yun ("Wind and Cloud") weather observation system, four of which had been launched as of November 2010. Capable of 3-D atmospheric detection as well as visible and infrared services, Feng Yun provides China with ultra-sophisticated tools for monitoring disasters, global climate change, precipitation, erosion, and a host of other phenomena that directly impact the Chinese economy and population.

The Medium- and Long-Term Plan for Scientific and Technological Development issued by the Chinese State Council for the period 2006–20 sets ambitious goals in continued progress to China's autonomy in high-technology, especially in the space sector. It stipulates that at least 2.5 percent of GDP be invested in research and development, that dependency on foreign technology not exceed 30 percent, and that science and technology constitute at least 60 percent of the Chinese economy.[64] Clearly, the space program and its associated technologies have taken a front seat in Chinese priorities, somewhat reminiscent of the policy prioritization toward space that took place in the United States in the 1960s during the Apollo program.

China's satellite launches originate from one of three space ports: Jiuquan, 160 kilometers south of the Mongolian border in the Gobi desert of Inner Mongolia (originally located there to bring Moscow within striking range of China's nuclear-tipped missiles); Xichang Satellite Launch Center, in Sichuan Province (the most modern); and Taiyuan, 500 kilometers southwest of Beijing. Ground was broken in 2009 on a fourth site, the Wenchang Satellite Launch Center, on Hainan Island, which is slated to begin operations by 2015. Once it is operational, most GEO launch missions will be relocated to Wenchang, and Xichang will become the backup launch site.[65] Hainan Island's location only 19 degrees north of the equator makes it ideal for future manned missions, since heavier payloads can be launched more easily and cheaply from lower latitudes.

In the 1990s China aggressively sought out international partnerships in space activities for what it declared would be the "the peaceful development of outer space." By 2006, China had signed 16 agreements with 13 separate countries, and initiated space industry production cooperation with more than 40 countries and agencies, including Argentina, Brazil, Canada, France, Malaysia, Pakistan, Russia, Ukraine, and the ESA.[66] In addition, China has signed cooperative memoranda with the space organizations of India and the United Kingdom. One of China's most celebrated collaborations has been with Brazil, which is an archetypical model of post-Cold War South–South cooperation. This collaboration resulted in the CBERS satellite series (I and II), produced in cooperation with the Brazilian Space Agency. CBERS and its successors have been used by both Brazil and China to track deforestation and other geographic phenomena.

While the space program has done wonders for China's image as a burgeoning space leader and for improving its working relationship with a number of states, the program's domestic importance as a facet of the country's defense strategy has been of paramount importance to the Chinese leadership. Just prior to assuming China's leadership, Jiang Zemin publicly argued at the 14th Party Congress in October 1992 that the success of national security would be predicated on future economic growth, and therefore military areas would necessarily be subordinated—for the time being—to the construction of the national economy. Once established, the economic foundations would provide the military all its necessities.[67] Once Jiang was in office, the current Chinese space policy emerged with his decree that the Chinese military would focus its efforts toward developing more high-tech approaches to national security in accordance with China's adoption of a "great-power mentality" (*daguo xintai*), which had evolved in response to the country's emergence as an economic power.[68]

This approach outlines the Chinese space policy succinctly: grow rich and only then grow militarily strong. High-ranking officers of the People's Liberation Army (PLA) openly commented that this approach was meant to counter the military superiority of China's perceived and potential adversaries: the United States, Japan, Vietnam, India, and Russia.[69] The State Council's 1998 defense white paper, entitled "China's National Defense," explicitly called attention to the need to offset US global hegemony, as well as to regional security matters such as US support for Taiwan and the long-standing US–Japanese defense relationship.[70]

A subsequent Chinese white paper, in November 2000, further outlined the space program's support of these intentions, though in strikingly less confrontational terms. Entitled *National Long- and Medium-Term Program for Science and Technology Development: 2000–2020*, the paper delineated the space program's three primary goals as (1) space exploration; (2) space applications; and (3) economic development.[71] In addition, the importance of science, education, and social progress was stressed as being integral to the country's future prosperity and security. To this end, China's space program today is primarily focused on the country's national development strategy, which has prioritized economic growth over other hard-power concerns, including many, though not all, military applications. Though 2035, China plans to launch upwards of 30 satellites with science and practical applications, such as one program that will produce wheat seeds in space that purportedly have superior yields after being subjected to high radiation in orbit.[72]

Nonetheless, in 2000 China also launched its first military communications satellite, which was the first in an eventual system of space-based

C4ISR (Command, Control, Communications, Computers, Intelligence, Surveillance, and Reconnaissance) capabilities. Also launched in 2000 were the *Jianbing-3* and the *Beidou 1A*. Built by the Chinese Academy of Space Technology, the *Jianbing-3* is a high-resolution remote-sensing satellite, which officially is intended for territorial survey, disaster monitoring, and space science. However, Western intelligence has identified it as the first Chinese military satellite able to provide the PLA with real-time satellite images that can be used for targeting down to five meters in accuracy.[73] The *Beidou 1A* was the first in a series of navigational satellites that will provide China with an autonomous global positioning system by 2015.[74] In pursuing these objectives, the overriding goal has been to augment China's autonomy and national sovereignty, with implicit aspirations to check US hegemony in the region, to stand up to the region's largest challengers, India and Japan, and most importantly, to challenge the US ability to dominate space.[75] The PLA utilizes space-based assets to support its operations in communications, global positioning, and reconnaissance, though by comparison with the United States, Chinese capabilities are still rudimentary.

Though it has been argued that China's manned program is simply a Trojan horse for the Chinese military's use of space, the truth is probably more complex. While one probable justification for the manned program is its appeal to nationalist sentiments, especially in the wake of the Tiananmen massacre, the need to address the ultra-high technology learning curve presents a more complete answer. As with the US experience, the dedication of massive scientific and technological resources to putting humans in space generates much human capital through greater educational attainment as well as technological advances and spinoffs. This approach results in spillover effects into many other sectors of the economy and society in general, including internal and external prestige-building.

China issued yet another white paper (a revised version of the 2000 paper) entitled "China's Space Activities in 2006," which contained the country's five-year plan for space activities.[76] Paralleling past US and Soviet declarations, China publicly stressed its commitment to promoting the peaceful uses of space, its cooperation with other space-capable states, and its opposition to the weaponization of space. The paper asserts that the space industry has always been as an integral part of China's comprehensive development strategy and puts heavy emphasis on the domestic policies behind China's space program, specifically areas such as the revitalization of the country through science and education, the development of an autonomous space science sector, and expanding the socioeconomic benefits of space activities. Nonetheless, more ambitious, arguably prestige-oriented missions, such as lunar exploration, are also listed as goals.[77] Longer-term goals that are highlighted include the establishment of a complete satellite industry (manufacturing, launching services, ground equipment, and operational services), supporting space research centers, and fostering the next generation of space scientists and engineers.[78] However, current trends in the Chinese space program suggest a more realist approach, emphasizing hard- as well as soft-power elements of space capabilities.

On 11 January 2007, China became the third country (after Russia and US) to successfully demonstrate the ability to destroy a satellite with an anti-satellite (ASAT) missile. After three previous failed attempts, China destroyed by kinetic kill one of its own derelict weather satellites (*Feng Yun 1-C*), orbiting at an altitude of 865 kilometers, with a precision that even earlier Soviet/Russian tests had never achieved (and in the process creating the largest orbital debris field in the history of space programs).[79] This feat has been noted as potentially having a greater impact on the future of international relations than the much more visible US invasion of Iraq in 2003.[80] It was furthermore interpreted by some observers in the region as the first direct post-Cold War attempt by China to influence the United States on trade and matters of regional hegemony, especially the 2009 India–US strategic partnership and their associated nuclear agreement. Most of all, China's demonstration of the mastery of this space application was meant to imply that, if necessary, China could eliminate essential US military and reconnaissance satellites, despite official Chinese statements to the contrary.[81] On the other hand, a contrary notion asserts that the test was a testament to China's large, confused bureaucracy not overseeing itself well enough (i.e., one branch of the government did not know what the other was planning).[82]

Near the end of 2007 China added another feather to its cap by becoming the fourth space-faring organization (following the US, the USSR, and the ESA) to put a spacecraft into orbit around the moon. The *Chang'e 1* mission (named for the Chinese goddess of the moon) provided the most accurate and highest resolution three-dimensional maps yet created of the moon's surface, including the dark side.[83] This accomplishment not only gained China praise from the world scientific community but was also a symbol of the country's progress in mastering the tracking, telemetry, and control technologies necessary for deep space probes.[84] The more sophisticated successor *Chang'e 2* was launched in October 2010, and carried out experiments from a low (100-kilometer) lunar orbit, including high-resolution mapping of Sinus Iridium, the proposed landing site of China's planned moon lander. The two probes constitute the first step in China's long-term lunar exploration program. The next phase will feature *Chang'e 3*, a six-wheeled lunar lander scheduled for launch in 2013 that will explore the moon's surface for three months, and *Chang'e 4*, a lunar sample return mission slated for 2017.[85]

The message was clear. In a single eventful year, China announced that it had arrived as *the* emerging space power. The Chinese ASAT test was carried out in part to exhibit the country's growing space capabilities, but also to throw down the gauntlet in reaction to the perceived US weaponization of space, in an attempt to even the playing field vis-à-vis a still technologically superior United States. China's action was provocative enough to bring an American counter-demonstration in 2008 with the shooting down of a defunct US spy satellite, *USA-193*, with an ASAT launch from the missile cruiser *USS Lake Erie*. The Chinese space program has followed a policy of defensive preparations that is itself part of a grand strategy meant to respond, primarily, to the military potential of the United States via the classic Cold War tactic of displaying hard power in order to elicit negotiations in the near term.[86]

Knowing that direct confrontation with the US is not feasible given US dominance in conventional weaponry (and exceedingly difficult to contemplate given the economic interdependence of the two countries), Chinese planners instead bet that, in the event of a Sino-US conflict, the US could be hobbled by targeting the US military's almost complete reliance on space-based reconnaissance, global-positioning, and communication systems.[87] This was stated as much by China's Central Military Commission, which held that "mastery of outer space" was the sine qua non of future combat.[88] Accordingly, the current Chinese plan for the country's space program is not to merely be on par with more developed space powers, but to in fact surpass them and dominate the orbital pathways.[89]

Even with China's orbital saber-rattling, the near-term goals of its program, like those of other space powers before it, seem meant to demonstrate China's space capabilities and potential in order to increase the country's credentials and standing among world space powers.[90] Space programs, particularly those including manned space flight which China has now successfully done three times, yield considerable prestige, which in turn can be transformed into both domestic and international political power and influence. As Asia's contender for regional hegemony vis-à-vis Japan, China is naturally interested in such matters. For example, China has eagerly sought a role in the International Space Station (ISS), a task that would have confirmed its exalted status. However, space ambitions are still, more often than not, muddled by terrestrial geopolitics. Acceptance of China's space participation by the West, especially Washington, would be interpreted as tantamount to tacit approval of the Chinese communist system and the country's controversial domestic policies.

China's current space program budget is estimated to be around US$2.2 billion, though a direct comparison to other space programs' budgets is problematic because of currency conversion issues with the Chinese renminbi, labor wage differentials, and the fact that the Chinese space program is highly integrated with the military and thus subject to the secrecy that accompanies such an association.[91] The most recent development of international note was the 2007 Sino-Russian cooperative agreement that called for a Chinese Mars probe, the 110-kilogram *Ying-huo-1* (the ancient Chinese name for Mars), to accompany the Russian *Phobos-Grunt* mission to the red planet. Launched in November 2011, the *Yinghuo-1* was to have photographed Mars and studied its magnetic field and the solar wind. This was to have been China's first deep space probe before it was lost when the Russian *Zenit* launcher failed, leaving the probes stranded in Earth's orbit.

But the most ambitious of China's long-term goals is a proposed moon landing before 2025, which a principal Chinese scientist attests is "a reflection of a country's comprehensive national power and is significant for raising our international prestige and increasing our people's cohesion."[92] Another reason to go to the moon as suggested by some Chinese officials would be to eventually mine the moon's abundant helium-3, an element not present in great quantities on Earth, but which is a potent fuel for the next generation of nuclear reactors.[93]

Such a feat would also paint an indelible image of China as a world power, since it would beat the United States' return to the moon, based on current NASA projections.[94] Besides the need to develop and perfect much of the technology necessary for such a venture, the key stumbling block is the fact that China's next-generation *Long March 5* (CZ-5) launch vehicle will still only have one-fifth the lift capability of the US *Saturn V* that sent *Apollo 11* through *17* to the moon. Nonetheless, the pursuit of a moon program would logically foster growth and innovation in the Chinese aerospace industry and promote a culture of technology for the next generation of Chinese engineers and scientists. In the short term, Project 863–706 proposes a Chinese space plane, called *Shenlong* ("Divine Dragon"), which will reputedly employ a ramjet engine and a maglev launch facility.[95] This could, presumably, be a response to the recently tested US X-37B autonomous space plane and the HTV-3X ramjet plane, which China's official media have called potential threats to Chinese security.[96]

Lastly, the growth of China's space program is likely seen by the Chinese government as important source of legitimacy, especially in its effort to portray China as a non-democratic model of development that others might follow.[97] The space program also illustrates how China is molding its cooperation with non-Asian states to fulfill its long-term political and strategic goals, thus using space activities for both hard- and soft-power advantages, particularly among traditional allies of the West in the developing world. The evolution of the Chinese program is therefore particularly pertinent, since China is forecast to become the world's largest economy before 2020. Joining the path blazed before it by earlier space powers, China has incorporated space as a essential element of national security and socioeconomic development in its drive to become the global hegemon of the twenty-first century.

India: Shiva, Buddha, and Kali in space

With the exception of human space flight, in technological terms, the Indian space program is roughly comparable to the Chinese program. With over 40 years of experience in space activities, India has gradually developed its missile and space programs almost entirely through its own efforts. Nevertheless, any examination of the evolution of India's rise as a space actor runs headlong into a principal question underlying this study: why would a country as relatively poor as India spend hundreds of millions of dollars on a space program when it could use satellites and other space technology from Russia, Europe, or the United States? The answer is rooted in the main argument of this book—that the pursuit of space activity brings with it the assurance of state sovereignty and the promotion of national development.

Even after a generation of advancements in public health and education, and despite India's having the world's tenth largest economy by nominal GDP and the fourth largest economy in terms of purchasing power parity, an estimated 42 percent of India's population still falls below the international poverty line.[98] But, as will be demonstrated, for India the strong civil roots of space activities nonetheless help to ensure national sovereignty, because Indian policymakers conceptualize national security in a more comprehensive manner than many other EMSAs.[99] The Indian vision of national security extends beyond the traditional realist notions to include human development as an equal pillar of national policy.

Since achieving independence from Britain in 1947, India has followed a fiercely self-sufficient foreign policy. But Indian officials insist that, in contrast to other countries, the foundation and rationale of the country's purely civilian space program is rooted primarily in concerns of national development and less in national defense. From the beginning, India's space program was meant to be a showcase of the country's advanced technology achievements, which would be justification for its acceptance as an equal alongside the developed world, and is thus attributable to the country's irresistible desire for long-withheld global recognition as well as a need to stoke national pride. Lastly, the space program fulfills a special role in India's goal of using technology to further state-led socioeconomic development and national independence, which Indian scientist and former president A.P.J. Abdul Kalam warned was necessary to ensure India's sovereignty.[100]

Historically, India's scientific and technological capacity in the space technology sector has been more or less equal to that of China, but its economic capacity has not. For this reason, India's 40-year-old space program has addressed practical concerns of national development as well as the country's most pressing security needs. Long ago, India prioritized the practical uses of space science over prestige. In other words, India's space program was designed from its foundation to be a vehicle for socioeconomic gains that would contribute to an encompassing notion of strategic national development.[101] So even while touting the recent success of a lunar probe or a possible Indian mission to the moon, most of India's space efforts have been directed to more down-to-earth initiatives such as telemedicine through satellite uplinks, online educational courses, and satellite communication for rural connectivity.

In addition, enhanced remote-sensing capabilities have vastly improved water and soil management, meteorology, and agriculture, and have even led to the discovery of previously unknown lodes of precious metals. The agricultural benefits are especially noteworthy in a country where 60 percent of the population still lives in rural areas. For example, the implementation of remote-sensing applications has provided farmers with highly accurate maps of cropland, giving them the ability to manage fertilizer, pesticide, and water usage in terms of square meters, not square kilometers, and thus saving money and precious resources. As a result, the tracking of pests, water flows, and crop diseases has been markedly improved.[102] In response to critics of the cost of the space program for a developing country, Indian officials argue that the space program additionally benefits the state by being cost-effective in certain developmental areas, such as addressing land use questions via remote sensing in comparison to conventional, labor-intensive approaches.[103]

But the practical Indian space program has nonetheless been influenced by the missile and nuclear imperatives that have driven all such capable states, and despite the developmental rhetoric of the Indian government, security considerations are still firmly entrenched, though murky, in the Indian space program. For example, the same satellite imaging techniques used so successfully to improve Indian rice yields can be used equally well to observe and track the Pakistani and Chinese militaries. Additionally, the satellite-based communications system that now links rural areas across the subcontinent can also provide the Indian military with a satellite-based communications network *par excellence*. The dual-use potential of the technology and the frequently contentious geopolitics of South Asia practically demand India's use of the space program to benefit the country's national security, and an increasing chorus of Indian policymakers are signaling the need to develop more offensive capabilities, such as an Indian ASAT, because India "lives in a dangerous neighborhood."[104]

In contrast to China's more circuitous and highly politicized path into space, India's space program has been less influenced by, though not divorced from, geopolitical concerns, and has gone through just two stages of development. The initial stage of some 20 years was primarily concerned with obtaining and/or developing the necessary technological infrastructure for space systems, such as sounding rockets. The second stage is characterized by India's devotion to the construction and utilization of high-capability flight systems. The progress from the first to the second stages has been focused on the practical outcomes of the space program.

However, India's focus on the practical socioeconomic applications of its space program should not be misconstrued to suggest that military interests have been completely eschewed or downplayed. Since gaining independence, India has sought to guarantee its sovereignty through a multitude of means, even if the consequences were not always palatable regionally or globally. Perhaps more than any other developing world space actor, India has aggressively pursued a complete and complementary package of missile, nuclear, and space technologies. In addition, India's quest for a presence in space serves the role of ensuring India's place in the emerging Asian space race that also includes Japan, China, Indonesia, North Korea, and South Korea.[105] What should be understood clearly is that, by design, India's space program is intended to serve its needs equally in defense, international relations, and socioeconomic development.

Like China's, India's space program has been mostly domestic in origin. India is no stranger to rocketry, and has consistently demonstrated a high level of technological competence in most areas of science and engineering. Historically, this was perhaps best demonstrated by Tipu Sultan's victory over British troops using rockets at the Battle of Guntur in 1780. Also like China's, India's space program was in part an offshoot of the defense-oriented policy that emerged from constant tensions with Pakistan, which itself begot India's ballistic and nuclear programs. After the country's bitter defeat in the 1962 Sino-Indian War and China's testing of its first nuclear weapon two years later, indigenous sources of weaponry, technology, and the socioeconomic development to drive them became the overriding policy goals of the Indian government.

In the same year as the Sino-Indian War, under the direction of Dr Vikram Sarabhai, the Indian National Committee for Space Research was set up to advise the government on space policy and to develop and test sounding rockets. This organization laid the foundations of India's space program. Initially, the program drew explicit distinctions between the civilian and military programs, in that the military arm developed liquid-fuel launchers because of its early reliance on Soviet-supplied SAM-2 rockets, while the civilian arm developed solid-fuel launchers.[106] From 1963 to 1975, India hosted hundreds of sounding rocket launches by the American, British, French, and Soviet scientists at the Thumba Range in northwestern India. Amidst this atmosphere of international cooperation, India's first home-grown sounding rocket, *Rohini-75*, was launched in 1967 (less than a month after China tested its first hydrogen bomb). Concurrently, Indian scientists began experimentation with US-supplied Scout rockets.

Near the end of this so-called observation period, India's current space agency, the Indian Space Research Organization (ISRO), was founded in

1969 as a department under the Atomic Energy Commission (AEC) headed by Sarabhai. With over 17,000 employees, the ISRO's official mission is the research, design, and production of Indian satellites, launch vehicles, and the necessary ground tracking systems. Highly telling is the institutional placement of the ISRO as an arm of the Department of Space, which itself is a branch of the National Natural Resources Management System. The ISRO answers to the prime minister. Importantly, military space programs are officially under a separate agency, the Defense Research and Development Organization (DRDO). The civilian space program has been purposely located within these agencies to highlight its developmental importance, in part because Sarabhai was against any form of space weaponization and advocated India's emphasis on excellence in advanced technologies to address the country's socioeconomic problems. Nonetheless, the prominent role of the space program as an icon of Indian nationalism and progress should not be underestimated given the country's long history of colonial subjugation.

India's push for self-reliance in its space and nuclear programs was, in part, a natural reaction to its recent colonial past under the British, but was also a practical response to the geopolitical realities of the day. Fearing India's non-committal stance during the Cold War, the US had blocked the export of supercomputer technology to India and, in a testament of Cold War détente, had convinced the Soviet Union not to sell India advanced cryogenic rocket engines. Isolated but undaunted, India forged ahead on its own, determined to produce what it needed without outside help.

Following China's 1970 launch of its first satellite, the ISRO established a 10-year nuclear-space program, called the "Sarabhai Profile," an ambitious plan that called for a self-reliant nuclear program and an advanced space program that would help develop missile delivery systems for both civilian and military purposes.[107] The plan called for the rapid development of the country's science and technology sectors as well as agriculture to ensure economic prosperity and "to ensure [India's] security in the world."[108]

The first successful step in India's space program came in 1975 when India's first indigenously produced satellite, *Aryabhata* (named after a celebrated Indian astronomer), was launched from the Baikonur Cosmodrome in the Soviet Union. The choice of launch site was ostensibly a testament to India's non-aligned status but was also to forestall possible US attempts to further influence the trajectory of the space program. For the rest of the decade, Indian scientists produced a series of small Earth-observation and communications satellites, largely in the quest to gain experience, refine engineering techniques, and confront the inevitable challenges of space flight.

In 1974, together with its push into space, India tested its first nuclear device, code-named "Smiling Buddha." Early nuclear technologies had been acquired by India from the United States in the 1950s, ironically as part of the Eisenhower Administration's "Atoms for Peace" program, which advocated the peaceful uses of nuclear energy.[109] In India's case, the original nuclear power energy production scheme was converted into what Indian officials called a "peaceful nuclear explosion."[110] However, its foreign policy purpose was obvious in that it signaled India's coming of age as a nuclear power with other associated capabilities on its horizons.

The link between India's space launch advances and its ballistic missile capabilities is patently clear, as the ISRO's activities support military missile programs through shared research, development, and production facilities.[111] India pressed forward to perfect solid fuels and guidance systems for its missiles in order to carry its new nuclear capacity, publicly trumpeting its accomplishments toward perennial rivals Pakistan and China.[112] The importance of this tactic was twofold: it indicated India's military potential while simultaneously demonstrating the country's credibility as a scientific leader of the Cold War Non-Aligned Movement, in which India was prominent. India's missile development to accompany its nuclear arsenal accelerated under the direction of A.P.J. Abdul Kalam. In 1980, India became the sixth country to build a satellite for geosynchronous orbit (launched by the ESA), but that same year also saw the first successful flight of its Satellite Launch Vehicle (SLV-3), launched from the Sriharikota Island launch site about 250 kilometers east of Bangalore on the southeastern coast of the state of Andhra Pradesh. This marked the beginning of India's autonomous space program, and the inauguration of the country's quest to develop an ICBM based on the SLV-3.

Shortly thereafter, in April 1982, India began the process of launching its own constellation of 21 communication satellites (the first four were built in the United States): the Indian National Satellite (INSAT), which had applications for a variety of transportation and personal communications, and the Indian Remote Sensing (IRS) satellite series, which provides visual and infrared reconnaissance for agricultural, ecological, geological, and cartography applications. Placed in polar sun-synchronous orbit, as of 2010 there were eight different satellites operational in this series. Weighing nearly one metric ton each, these two different series of satellites, launched from the Soviet Union (later Russia), propelled India into a very small club of states capable of autonomously directing its own social and economic development via space applications. Also important to India's developmental goals was the *Rohini-3* communications satellite launched in August 1983, which extended television coverage from 20 percent to 70 percent of the country's population (today the figure stands at 90 percent).

India's swift rise as a potential space power shook Washington and raised suspicions that India's growing missile technology had a less-than-benign intent. The 1987 Missile Technology Control Regime (MTCR), initially signed by most of the largest space powers, was crafted to slow or stop many of the ballistic missile programs in the developing world. Because of the dual-use nature of space technology, India's space program fell under this US-led technology export ban. Undeterred, India tested its first intermediate-range ballistic missile, the *Agni* (named after the Hindu god of fire), in 1989. But when in 1991 Russia (not a signatory of the MTCR) agreed to sell India much more powerful cryogenic engines, the US protested. A compromise was reached in which Russia sold India only some engines with important technologies, which forced India to complete the engines on its own, and left Indian policymakers with bitter resentment toward the United States.[113] Unexpectedly, however, this ban had the effect of unifying the country's academic and private enterprise sectors to undertake space-related projects that might not have been pursued otherwise. Thus, in effect, the MTCR's prohibitions actually promoted improvements in missile and, subsequently, space technology in India.[114] On 20 September 1993, India performed its first polar launch with its indigenous *Polar Satellite Launch Vehicle* (PSLV).

Besides focusing on internal development, the ISRO has sought to make India one of the world's preferred launch centers. After India's economic liberalization program of the 1990s, the goals of the ISRO were retooled toward the international market, and with its upgraded capabilities, India has followed China's example to actively court the international launch market, which has revenues of around US$3 billion annually.[115] Again mirroring China, the ISRO set up a marketing and business division in 1992 called the Antrix Corporation, which serves as the marketing arm of the country's space agency. The main launch vehicle for this program is India's Geosynchronous Satellite Launch Vehicle (GSLV), a three-stage, 400-metric-ton launcher first tested in April 2001 whose development was to free India from dependence on foreign launchers. The GSLV will enable India to independently sustain its communications and navigation satellite network.

The crowning moment in the recent history of the India space program was the successful launch in 2008 of the *Chandrayaan-1* probe to the moon using a PSLV-XL launch platform. Loaded with indigenous instruments as well as some provided by the ESA and NASA, the probe snapped over 70,000 images of the moon. *Chandrayaan-1* also verified the existence of water on the moon using an impact probe, thereby boosting India's credibility as a serious space actor. While it operated at the relative bargain cost of US$97 million, the *Chandrayaan-1*'s primary mission of lunar mapping via high resolution remote-sensing equipment was nonetheless of limited scientific value since about 97 percent of the moon's surface had been already mapped by US, Russian, and most recently Japanese probes. Additional probes to the moon are scheduled for 2014 (*Chandrayaan-2*) and 2015, and will include a wheeled rover that conduct chemical analyses of lunar soil. These two projects will be carried out in cooperation with the Russian Federal Space Agency and will ferry a Russian rover to the lunar surface.[116]

The immediate importance of the *Chandrayaan-1* mission was that it served as a practice opportunity for more complex projects in the future and that it seemingly vindicated the country's proportionally large investment in space to its still largely poor population as well as to the skeptics within the Indian government. Equally important was the boost it provided to India's national prestige, the enhancement of the country's image as a technology center, and the demonstration of the potential of India's ever-more sophisticated space technology. Lastly, *Chandrayaan-1* is most certainly India's contribution to the emerging Asian space race as well as the most visible symbol of the country's deterrent capabilities, given that India's space launchers are the basis for its ballistic missiles, and Indian strategists admittedly measure their security efforts in relation to the perceived threat from China.[117]

The vision of the ISRO is ambitious and in line with the country's ascension as a global economic player. India's indigenously produced GSLV is expected to be in service by the end of 2012 to hoist an ever-increasing armada of Indian and foreign-paying satellites into orbit. The eventual goal is to be able to lift 10-ton satellites into low Earth orbit.[118] The GSLV launcher series will also be the backbone of India's plan to achieve its first manned orbital mission by 2016, a Mars probe by 2020, and a proposed manned mission to the moon in 2025.[119] India is also in the design stages of developing of a reusable, ramjet-powered space plane called *Avatar*, which is a joint project between the Defense Research and Development Organization and the Space Research Organization.[120]

An important first step toward putting an Indian in space occurred in February 2009 when the Indian government approved the manned program and almost US$3 billion for its development. While publicly portrayed as merely an extension of India's growing technological prowess, the manned program is probably a reaction to China's achievements in manned flights, and will also advance the country's expertise in advanced missile systems and aid in the mastery of manned space flight technologies. While derided by some detractors as a misdirection of Vikram Sarabhai's original guiding vision for the Indian space program as a vehicle for national development instead of grandiose displays such as manned flight and space probes, India has nonetheless emerged as an emerging space actor for whom manned space flight will be a normal and logical step in demonstrating its abilities and national aspirations as well as furthering its technological frontiers.

The importance of space policy in India today is furthermore demonstrated in its increased overall funding. In 2003, the ISRO's annual budget was a relatively modest US$200 million, while by the end of the decade the budget had soared to over US$1 billion (though even considering the five-fold increase, funding for India's space program is still a meager 3 percent of that enjoyed by its American counterpart, NASA). So, like China, India has become essentially self-sufficient in its space program and, likewise, seeks to use its abilities for a host of purposes in the strategic and socio-economic realms. But even with its goal of self-sufficiency, India has continued to purposely engage in cooperative agreements with the Big Three space powers, the United States, Russia, and the European Space Agency, both to bolster its image as a space power and to reaffirm its traditional non-aligned status. Given India's recent ascent as an economic power, India's leadership envisions the country as a world leader in space technology, especially in the application of these technologies to social and economic development. In addition, the Indian space program plays an integral role in restoring greatness to a society that is the heir to a great civilization whose accomplishments in science stretch back millennia.

According to the ISRO's official mission statement, the country's space program is completely oriented toward applications intended for national development. Indeed, many of India's achievements in space have been designed to target improvements in areas such as water management, cell phone communications, telemedicine links with remote corners of the country, and even satellite television to the masses. The country's remote-sensing satellites have been vigorously employed to track erosion, crop yields, and land usage.[121] But despite all its technical and developmental accomplishments, the Indian space program still fulfills the primary goal that such programs have had since the beginning: it is a powerful and visible symbol of India's aspiration to greatness, it builds confidence and self-esteem in the country's population, and it gains India international prestige and influence. Even the election in 2002 of A.P.J. Abdul Kalam as India's eleventh president had substantive and symbolic import, as he had previously worked as a scientist for the ISRO for almost two decades as well as being a leading figure in the development of the Indian missile and nuclear programs.

Brazil's samba into space

Though typically more widely recognized to the outside world for *carnaval* and *futebol*, Brazil has for the past half-century quietly pursued a dedicated program of technological development in the areas of defense and space technologies. Even the presence of a Brazilian astronaut aboard the International Space Station (ISS) in March 2006 is merely the latest step in the country's journey to establish a successful space program that will bolster its standing as a rising world power. The development of Brazil's indigenous space program has evolved as a natural extension of the country's long-running strategy to establish itself, at a minimum, as Latin America's hegemon, with respect not only to South America but to the entire South Atlantic Ocean region. In the long term, a successful Brazilian space program would also provide additional justification for the country's aspiration to a permanent seat on the UN Security Council, a change which is now supported by Britain, a more advanced space actor.[122]

With 203 million people, Brazil is Latin America's most populous country as well as its largest. Brazil also has the world's seventh-largest economy by nominal GDP and the eighth-largest in terms of purchasing power, with an economy bigger than all other South American countries combined. Moreover, Brazil stands alone in Latin America in terms of its space-related capabilities and ambitions. While a number of Latin American countries have cultivated and engage in some limited space-related research and basic launch capabilities, these endeavors have been largely confined to the development of satellites and/or related aerospace technology, and these efforts are frequently dependent upon outside partners for technological and/or financial assistance. What is more, practically all these states have been dependent on other states for launch capability. By contrast, while Brazil has collaborated with, and at times even depended on, more advanced space actors to help build its space program, it has now begun to approach the point of achieving independent launch capability with space projects that are considerably more sophisticated, diversified, and advanced than those of most other emerging space actors.[123]

Brazil's increasing emphasis on the development of its space capabilities has become a vital component of the country's national security and socioeconomic development strategies. The Brazilian government's rationale for its space endeavors is unambiguously expressed as being strategic for the sovereign development of Brazil … only those countries that master space technology will have the autonomy to develop global evolution scenarios, which consider both the impact of human action, as well as of natural phenomena. These countries will be able to state their positions and hold their ground at diplomatic negotiating tables.[124] Brazil has always considered itself *primus inter pares* in the developing world and, clearly, Brazil has made the same assumptions as earlier space actors—that space is simply the newest arena in which a country must exercise its national power to ensure its sovereignty and further its national interests.

Specifically, the pursuit of an independent space program falls in line with the trajectory of Brazil's foreign policy in the twenty-first century, a policy that has become strongly proactive. Brazil's forward-looking disposition regarding space activities has been spurred by the prestige associated with mastering nuclear technology, a desire to win a permanent seat at the UN Security Council, and the goal of attaining leadership regionally and sharing it globally. Even more specifically, the need to protect Brazil's vast resources, in particular the newly discovered Tupi and Jupiter offshore oil and natural gas fields (purported to be some of the world's largest), has raised questions about the readiness of the Brazilian military to protect them. In response, Brazil has moved boldly to initiate ever larger programs in military modernization and nuclear technology—whether for a submarine reactor, uranium enrichment, or, potentially, nuclear weapons—which add to the country's perception of its security as well as its prestige in international fora. The space program is the third leg in the Brazilian national security triad.

Brazil's current space policy can be summarized by three general goals: (1) to exert sovereignty over its vast, rich, but thinly populated geographic interior; (2) to develop economically and militarily so as to obtain a presumably deserved regional leadership position; (3) to eventually receive recognition as a world power.[125] In territorial integrity, Brazil's status as Latin America's largest country (and the world's fifth largest) makes it a natural, though not unchallenged, hegemon. But with upwards of 80 percent of the country's 191 million people living within 400 kilometers of the Atlantic coast, a considerable portion of the country is underpopulated, averaging only 18 persons per square kilometer. Brazil has adopted a number of other policy initiatives over the years to extend and strengthen its territorial integrity. To better understand the implications of Brazil's foray into space activities and the role its space program plays in the country's national security and developmental strategies, it is useful to contextualize the program within the evolution of the country's broader national security programs and developmental needs.

First championed by President Castelo Branco in 1966, Operação Amazónia (Operation Amazon) was a program to encourage migration into the interior, and in 1970 the Plano de Integração Nacional (National Integration Plan) was begun to ensure national control of Brazil's vast interior through road construction, population resettlement, and agricultural subsidization. The same year, as a result of the so-called "Lobster War" with France (a dispute over fishing rights), Brazil unilaterally extended its territorial waters to 360 kilometers offshore.[126] Lastly, in 1984 Brazil extended its reach by declaring a "zone of interest" in Antarctica (making it the third Latin American state after Argentina and Chile to do so), though as a signatory to the Antarctic Treaty System of 1959 it has not formally made territorial claims. This move was part of what Brazilian geopolitical strategists called *defrontação*, a plan that espoused a greater South Atlantic presence for the country.[127]

A long-standing theme in Brazil's quest to be recognized as a world power has been the effective utilization of its extensive natural resources for economic development, which has manifested itself in a number of ways. From moving the capital from Rio de Janeiro to Brasília in 1960 to various massive hydroelectric projects and extensive agricultural endeavors, the boldness of Brazil's undertakings has paralleled the country's ambitions to be recognized as a rising power.

It is important at this juncture to note that in addition to the tangible factors outlined, another, more subjective, matter should be considered as providing some level of justification for the country's expansive development projects and security agenda: this is Brazilian society's perennial notion of *grandeza* ("greatness"). Despite the fact that almost one-third of Brazilians live in poverty, Brazil's people have traditionally seen their country as a natural regional and potential world power. This concept of *grandeza* goes a long way toward understanding the logic of Brazil's national development and defense priorities. This perception of destiny is bolstered by Brazil's geostrategic location along the sea lanes of communication in the Atlantic, from the equator to Antarctica.

This notion of regional hegemony had been traditionally resisted by Brazil's southern neighbor and perennial rival, Argentina, which has likewise pursued a variety of projects in nuclear power and weapons, ballistic missiles, and eventually, space programs (see Chapter 5). But after Argentina's defeat in the 1982 Falklands/Malvinas War, Brazilian geostrategists felt that the time was right for Brazil to fill the power vacuum. This plan would build upon the goals of former President Juscelino Kubitschek, who declared that Brazil would produce "fifty years of development in five [years]." From the late 1950s to the mid-1970s, during a period known as O Grande Brasil (Great Brazil), the Brazilian government undertook a series of daunting projects, which read like a list of engineering hyperbole: the world's longest bridge, the world's largest hydroelectric dam, the Trans-Amazon Highway, and plans for a network of up to 10 nuclear power plants (to be completed in cooperation with West Germany).[128]

The beginning of the expansion of Brazil's geopolitical consciousness is found in the early twentieth century up through World War II. During this time period, Brazil's defense spending spiraled upwards in response to perceived challenges by Argentina, and to establish Brazilian hegemony in South America, particularly because of Argentina's covert involvement in the Chaco War (1932–35) and the Argentine military's pro-Axis sympathies during the early 1940s. Brazil's eventual entrance into the Italian campaign during World War II with the Força Expedicionária Brasileira (Brazilian Expeditionary Force) marked a turning point. Although its contribution to the overall war effort was relatively minor, Brazil was one of only two Latin American countries to actively participate in the war (the other being Mexico, with its contribution of an air fighter squadron—Escuadrón 201—in the Pacific theater). Brazil's active participation in world affairs was perceived early on by Brazilian strategists as essential to the country's aspirations to be taken seriously as an aspiring world power.

The creation of a space program fit logically into these grandly ambitious designs, and successive military governments of Brazil (1964–85) predicted with all confidence that the country would join the world's space powers, launching Brazilian-made satellites on Brazilian-made rockets. It was also assumed that the space program would lead a traditionally inward-looking country toward some degree of technological independence in diverse sectors, such as informatics, arms industries, nuclear energy, and satellite technologies. Brazil's pursuit of this recognition led the county to begin development of an independent rocket and nuclear energy program in the late 1950s.

During the administration of President Juscelino Kubitschek (1956–61), Brazil began to develop an indigenous nuclear energy and weapons program, partly in response to a similar program in Argentina. These nuclear ambitions were accelerated by subsequent military governments, which pursued a variety of uranium enrichment approaches for all the military services.[129] The plan went so far as to construct a 300-meter-deep shaft in the northern state of Pará for never-completed underground nuclear tests. Thus the genesis of Brazil's modern space program can be traced to the development of the country's nuclear enrichment and ballistic missile programs during the years of military dictatorship.

The precursor to the formalization of Brazil's space program occurred during the International Geophysical Year of 1957–58 when, through the US Naval Research Laboratory, Brazil set up US monitoring station in São José del Campos near São Paulo to receive data from the US *Vanguard* satellites. In 1961, President Jânio Quadros established the Grupo de Organização da Comissão Nacional de Atividades Espaciais (Organizational Group of the National Commission of Space Activities—GOCNAE) to examine the country's needs in order to develop a viable space program.[130] The establishment of GOCNAE made Brazil one of the world's first developing countries to formally sponsor space activities, but it is noteworthy that the Brazilian military was deeply involved in the space program from the beginning.

One of the military government's concurrent priorities was the development of an indigenous ballistic missile program, and by 1965, Brazil was launching sounding rockets from its newly constructed Centro de Lançamento Barreira do Inferno (Barrier of Hell Launch Center) in the state of Rio Grande do Norte. The first launches from this new site were Brazil's contribution to the International Quiet Sun Years of 1964–65 (a complementary successor program to the IGY). The country's domestic meteorological program began the following year. More than 2,000 successful launches have been carried out from this site.[131] During this period, Brazilian strategists began to envision a space program's contribution to national security, based upon three general strategy areas: resource management, economic and state development, and defense and territorial integrity.

Over the next 30 years, Brazil would spend about US$1.5 billion toward improved ballistic technology, even creating university engineering and physics programs to sustain the project. Due in large part to these advances, Brazil became one of the original signatories of the 1967 Outer Space Treaty (OST), which, among other things, forbade the placement of weaponry in orbit. Though 91 countries signed the OST, Brazil was one of the few signatories from the developing world that actually had both the ambition and potential to develop a missile and space program that could eventually impinge on the treaty.

Even so, prompted by ongoing competition spurred by Argentina's ambitious *Cóndor II* ballistic missile program in the 1970s and early 1980s, Brazil put increased resources into improving defense-related technology, especially ballistic missile technology. The program took a decidedly security-oriented turn as the country's military industries were expanded significantly in the 1970s. The first bureau for space-related technologies opened in 1969, called the Instituto das Atividades Espaciais (Institute of Space Activities—IAE).[132] The IAE was consolidated in 1971 into the Comisão Brasilera das Atividades Espaciais (Brazilian Commission of Space Activities—COBAE), under the Ministry of Aeronautics. Chaired by the head of Brazil's Armed Forces General Staff, the aim of the COBAE was unreservedly security-oriented as it sought to achieve Brazilian self-sufficiency in missile technology. The program was sufficiently successful to prompt the United States to enact a ban on the export of missile technology to Brazil, since the US had strong reservations about a ballistic-missile-armed, and potentially nuclear-armed, Brazil. As it turns out, this apprehension was not completely unfounded.

Brazil had begun its own research into nuclear fission as early as the 1930s. Brazil's nuclear research was initially facilitated by technology and uranium fuel transfers from the United States in the late 1940s, and the Brazilian navy took an active lead in stimulating national nuclear energy and electronics programs. The Consejo Nacional de Pesquisa (National Research Council—CNPq) was founded in 1951, largely through the efforts of the Brazilian navy, to consolidate state control over nuclear activities. The CNPq would later evolve into a broader instrument for the support of many areas of research and development. In 1953 Admiral Alvaro Alberto concluded a secret agreement with the West German Institute for Physics and Chemistry (headed by one of the former leaders of Hitler's nuclear weapons program) to buy three centrifuges for uranium enrichment.[133] The equipment was confiscated by occupying US forces, held for three years, and then released to Brazil after the US occupation of West Germany ended. In the meantime, Brazil had acquired two nuclear research reactors and uranium fuel from the United States in 1957 under the Atoms for Peace Program. However, successive Brazilian military governments felt increasingly hamstrung by restrictions imposed by Washington on technology transfers.[134] After the construction of its first nuclear energy plant in 1971 by the US company Westinghouse at a coastal site between Rio de Janeiro and São Paulo, the Brazilian government eventually decided to pursue a home-grown nuclear program, which reflected the government's aspiration to be independent, while retaining the option of integrating on equal terms with other large, powerful states.

For this reason, Brazil once more went outside Washington's purview and again approached West Germany, entering into an agreement in 1975 which was to provide up to eight nuclear reactors without International Atomic Energy Agency (IAEA) oversight. Though a signatory to the 1967 Treaty of Tlatelolco, which banned nuclear weapons in Latin America, Brazil's military government nonetheless felt a nuclear option was crucial to the country's long-term security plans: it allowed the country to begin to transfer nuclear technology into a covert program of uranium enrichment, code-named Solimões (aptly named after the beginning of the Amazon River in Brazil). The objective was to master all phases of nuclear energy production, including those with potential military applications.[135] Coupled with the country's growing missile program, this nuclear arrangement became the foundation for Brazil's defense program.

However, following the end of military rule, the nuclear weapons program was publicly repudiated by President Collor de Mello in 1990, and in 1997 Brazil entered into the Nuclear Non-Proliferation Treaty (NPT). Domestic legislation (Law 9112) was created to regulate the export of nuclear enrichment technology, which in part was seen as a necessary step to membership to the Missile Technology Control Regime (MTCR), which would allow the importation of foreign civilian space-related technologies.[136] Nonetheless, indications are that the Brazilian military continued to circumvent these intended controls and continued a surreptitious program of perfecting nuclear enrichment.[137] Though a signatory to the NPT, Brazil has continued to include nuclear power as part of its strategic plan, and Brazil has continued to actively pursue an accelerated program of nuclear energy. First declared in 1975, the official policy of Brazil is to be completely self-sufficient in uranium for nuclear electricity production by 2014, and still produce enough for export.[138]

In symbiotic growth with the nuclear and ballistic missile programs was the development of Brazil's defense industry, which produced a variety of affordable, top-quality armaments. Brazil's military industries grew dramatically during the 1970s, reversing the country's long-standing dependence on foreign military suppliers, and made Brazil an arms export and research leader among developing countries.[139] Brazil's defense industry reached sufficient capacity and quality that the country became one of the world's top exporters of small arms, radars, main battle tanks, missiles, and even nuclear and chemical weapons technologies, selling to at least 42 different countries throughout the world.[140] An extensive network of defense industries flourished, and by the mid-1980s Brazilian manufacturers produced 80 percent of the weapons used by the Brazilian armed forces.

Brazilian arms sales reached their zenith in 1989, and in the same year Brazil became the world's eleventh largest arms exporter, with over US$380 million in foreign sales.[141] During the 1980s, its largest regional market was the Middle East, to which Brazil sold roughly half of its arms, with nearly half of the US$1 billion in Brazilian arms transfers from 1985 to 1989 going to Iraq during the Iran–Iraq War.[142] One of the most successful and profitable Brazilian exports of this period was the Astros II multiple rocket launcher, produced by Avibrás Indústria Aeroespacial, a company that specializes in rocket, missile, aircraft, and telecommunication technology. At the same time, Avibrás was developing ballistic missiles for the Brazilian military with ranges of up to 1,000 kilometers. In June 1989, Avibrás entered into a joint venture with the Chinese Ministry of

Aeronautics and Aeronautics Industry to sell space technologies and launch facilities to developing countries.[143] These systems and many others created lasting relationships between civilian contractors and the Brazilian military, and fueled the growth of domestic high-tech suppliers. Brazil's military sales ambitions went beyond just delivery systems. From 1981 to 1982 Brazil secretly sold uranium dioxide (used in nuclear fuel rods) to Iraq without notifying the International Atomic Energy Agency (IAEA).[144]

One early measure of Brazil's success in missile technology was the fact that the country's technological capacity had grown so fast that it was one of only two developing countries (Argentina was the other) to be a signatory of the 1987 Missile Technology Control Regime (MTCR), an agreement that sought to control the proliferation of nuclear-capable ballistic missiles. The agreement did successfully delay Brazil's missile program, as the country's proposed missile development cooperation program with France was pressured into cancellation by the MTCR. After the fall of military rule in Brazil via the *abertura* (opening) in 1985, the defense industry fell into disarray; Brazil's arms exports shriveled by the early 1990s, with only US$3 million in annual sales, and its three largest arms manufacturers fell into bankruptcy. As a result, the totality of Brazil's ballistic and missile programs were transferred into civilian hands by 1994. While Brazil's military-industrial complex had peaked and waned, its space program progressed steadily.

In 1981, the military-run COBAE became the Missão Espacial Completa Brasileira (Brazilian Complete Space Mission—MECB), which was charged with addressing a broader range of both national security and developmental concerns that reflected Brazil's acknowledgement of more complex national and international realities. The program was endowed with a relatively generous US$1 billion budget, and its stated objectives were expanded to include a broader set of national priorities: (1) seek out and monitor natural resources; (2) map the Amazon region and track deforestation; (3) oversee agricultural activities; and (4) provide telecommunications.[145] In addition, Brazilian officials openly stated a desire to use the country's launch capability to make Brazil competitive in the international commercial space launch market, "including the military applications sector."[146]

To achieve this goal, construction began in 1982 on the Alcântara Launch Center on Brazil's northern Atlantic coast in the state of Maranhão, a tracking station at Cuiaba in the western state of Mato Grosso, and a mission control center in São Paulo. Costing US$300 million and built on 62,000 hectares expropriated by the Brazilian air force from the local inhabitants, Alcântara, at less than two degrees latitude south, is the world's closest launch facility to the equator. This location makes it the best launch site in the world in terms of fuel efficiency, load capacity, and downrange safety because of its wide downrange north and east launch azimuth. Launches from Alcântara have approximately 25 percent greater launch energy to achieve orbit than those from any other launch site in the world. In addition, in the case of an accident, debris would fall into the South Atlantic Ocean.[147]

Operated by the Brazilian military, Alcântara houses its own meteorological, telemetry, and vehicle assembly operations. Brazil hoped to benefit from the facility by learning the cutting-edge technology that would eventually help to create the country's own satellite industry. Alcântara became operational on 19 August 1994 with launch pads available for the Sonda sounding rockets, meteorological rockets, and other science projects, as well as for Brazil's homegrown VLS booster. Foreign customers were courted immediately, with France testing its *Ongoron I* and *Ongoron II* and NASA launching a handful of its Nike Orion sounding rockets from Alcântara during its first year in operation.

The quest to be a space-faring state was, at one time, part and parcel of the tug-of-war between civilian and military supremacy in Brazil. The country's first truly civilian space agency, the Agência Espacial Brasileira (AEB), was created in 1994, whereas the majority of the earlier space program research had been under the unyielding control of the Brazilian military. But despite Herculean efforts, it was not always clear that Brazil would achieve its goal of an autonomous space program. Some observers doubted the prospects: a 1993 Rand Corporation study concluded that Brazil's space ambitions, like those of many developing countries, were not economically viable.[148] However, this assessment was tendered prior to Brazil's rise as an economic power, which has since provided the country with the resources to realize projects to address its perceived national security and developmental needs. Brazil increased joint space program efforts with China and Russia to circumvent technology transfers denied to it by the United States. Though the US at last waived its objections to Russian technology sales to Brazil, by 1996 concern was being voiced about Brazil's (re)acquisition of ICBM technology.[149]

The current AEB is administered under the National System for the Development of Space (SINDAE), the umbrella organization set up in 1996 to oversee all of Brazil's space activities. Though direct military oversight of space programs has been eliminated, collaboration with military-run research programs continues.[150] Nonetheless, the lion's share of AEB's budget is dedicated to remote sensing, which promotes Brazil's domestic space industry.[151] Brazil's indigenously produced launch vehicle, the *Veículo Lançador de Satélites* (Satellite Launch Vehicle—VLS), has had a rocky start. As a joint venture between the civilian AEB and the Brazilian Air Force, the VLS was originally envisioned as Brazil's answer to the European Space Agency's *Ariane-5*—a powerful and dependable launch vehicle for domestic and for-profit foreign satellite launches.

But the accidental explosion at Alcântara on 22 August 2003 of the VLS-1's first stage seemingly quashed Brazil's ambitions. The explosion killed 22 Brazilian engineers and technicians and reduced the launch pad to rubble. Attributed to insufficient funding and lax management, it was the third failed launch of the VLS-1 rocket (previous attempts were made in 1997 and 1999), which was designed to carry two satellites into orbit. The result was a more difficult job to attract suppliers, whose numbers dwindled by two-thirds.[152] Nonetheless, launch failures have been a fact of life for all major space-faring countries in the development of launch vehicles (for example, between 1957 and 1999 the US and Russia experienced a combined 345 launch failures or an average 10 percent failure rate[153]). Only 14 months later Brazil resolutely launched from Alcântara a smaller, 12-meter-tall VSB-30 rocket, which carried a microgravity experiment called Cajuana at an altitude of 100 kilometers; a second successful launch the same day achieved 259 kilometers in altitude. In 2007, 2009, and 2010, Brazil independently launched three rockets for the Operation Maricati project, each containing atmospheric tests. The last flight's experiment was recovered for further tests. So while achieving comparatively modest gains so far, these successes have put Brazil on the map as a space port of great potential importance.

In addition to pressing forward with its own launch program, Brazil has continued to build a reputation as a dependable partner with larger space actors, though at a cost. In 1997, at the invitation of the Clinton administration in the US, Brazil became the only developing country among a long list of developed space actors to contribute technology to the International Space Station. Besides being a gesture to an emerging space-faring developing country to join the program as part of NASA's quota for the ISS, the invitation was additionally a machination of the Clinton administration intended to favorably mold Brazil's space and nuclear program toward US interests.[154] Initially promising a US\$120 million contribution of flight equipment, Brazil was later forced to pare down its portion to US\$10 million because of the country's then-high and persistent foreign debt. Though smaller, this contribution does fit with Brazil's interest in space cooperation, a tactic that Brazilian officials undoubtedly expect will pay dividends in furthering Brazil's position as a rising world power.

Brazilian space cooperation has extended beyond its traditional relations within the western hemisphere. In July 1988 Brazil and China signed a protocol of cooperation in the development of the high-resolution remote-sensing satellites *CBERS-1* and *CBERS-2*. The cooperation was highly successful and has been praised as an example of South–South cooperation in technology.[155] China installed its own satellite control equipment at a ground station in Brazil. In October 2004 Brazil signed further agreements with China to develop the high-resolution *CBERS-2B* imaging satellite, which was launched in 2007 aboard a Long March 4B rocket from China's Taiyuan Satellite Launch Center. There is an agreement for the option of two more models extending to 2014. In a gesture of reciprocity, China has discussed the possibility of shipping Long March rockets to Brazil for launch from Alcântara.

While cooperative agreements with China have borne fruit, Brazil has also sought other partners for space cooperation. Beginning in 1998, Brazil approached erstwhile rival Argentina to cooperate in joint microgravity space experiments, which finally took place with success in 2007 with the launch of a Brazilian VBS-30 sounding rocket. The result was impressive enough that the Swedish Space Corporation purchased the VBS-30's rocket motor for use in its own sounding rockets for the ESA. In 2008 Brazil and Germany signed an agreement for a Multiple Application Synthetic Aperture Radar satellite that will possess night-vision capabilities for monitoring the often cloudy Amazon. And in 2010, Brazil entered into a trilateral agreement with India and South Africa that will produce two satellites—one for Earth observation and the other for space weather and climate studies.

The CBERS project bespeaks the particular importance that Brazil has put in recent years on the ability to monitor the deforestation of the Amazon region, which comprises about one-third of the country and approximately two-thirds of all the tropical forests on Earth. Though it had historically ignored illegal logging activities, in a striking broadening of the definition of national security, the Brazilian government declared the preservation of the Amazon rainforest a matter of national security.[156] Beginning in 1988, the Brazilian Science Ministry monitored deforestation via NASA Landsat satellite imagery. But with Brazil's first satellite in 1993—the *Data Collecting Satellite 1* (SCD1)—Brazil began to monitor the region through its own Program for the Estimation of Deforestation in the Brazilian Amazon (PRODES). The program has yielded impressive though sobering results. Fitted with a multispectral camera, CBERS satellites have revealed that the Amazon rainforest is disappearing twice as fast as previously estimated, releasing more than 100 million more tons of carbon dioxide into the atmosphere each year than had been previously calculated.[157] Necessarily, Brazil is now on the cutting edge of the study of carbon emissions from forest burning and cattle ranching, the two main sources of tropical deforestation around the world. Like an expanding number of developing countries, Brazil now uses its satellite applications to monitor changes in land usage, the ocean, and pollution, in the service of and protection of the national good.

The pinnacle of Brazil's satellite-based efforts to protect the Amazon is the SIVAM (System for the Vigilance of the Amazon) project. Originally envisaged in the 1990s as a system to monitor the five million square kilometers of Brazil's Amazon rainforest, it is regarded as one of the world's most elaborate environmental protection and law enforcement schemes. The system resulted from a joint endeavor between the US-based Raytheon Company and the Brazilian companies ASTECH and Embraer, the third-largest civilian aircraft manufacturer in the world. SIVAM became operational in 2004 and uses satellite imaging, airborne assets, and ground-based radars and control stations to monitor for illegal loggers, miners, and drug traffickers in the Amazonian region. However, some observers have argued that the project is merely a convenient cover for the Brazilian military's expansion into space-based reconnaissance of the region in order to protect it from foreign exploitation of Brazil's resources and the incursion of criminal elements from neighboring states.[158] Brazil also currently operates or plans a number of other Earth observation satellites dedicated to the protection of the Amazon, such as the *SSR*, which is a remote-sensing platform to monitor land use, and the *SCD-1*, an environmental data collection platform.

But despite a growing track record of operational successes, the Brazilian space program has struggled to overcome bureaucratic infighting, corruption, and the almost legendary misallocation of funds. Before 2003, 95 percent of the feeble US$10-million space program budget went to Embraer and only 0.5 percent found its way into the civilian space program.[159] But following the devastating Alcântara explosion, the Brazilian government adopted a completely different approach. In addition to opening up the program to outside scrutiny and advice, principally in the form of Russian advisors, the government drastically increased funding to US$100 million for the 2005 fiscal year. While this figure trailed India or China's annual space program budgets of US$300 million and US$1.8 billion respectively,[160] it nonetheless represents a 235 percent increase over 2003 outlays and is a clear sign of the importance that the Lula da Silva government put on the space program. In 2009 the Brazilian government upped the ante again and allocated US$343 million to the space program, finally putting its funding closer to parity with other first tier space actors.[161]

Along with its drive to develop autonomy in its launch systems, Brazil has been equally ambitious in seeking partners to bolster its capabilities and its image as a space partner. In October 2003 Brazil's INPE signed a joint venture with Ukraine for commercial missions, which was to launch a Ukrainian Tsyklon 4 medium-class rocket from the Alcântara launch site. However, the 2004–05 Orange Revolution in Ukraine upset those plans, though in 2009 a similar agreement was reached to carry out the launch in the 2012–14 period, which itself has been delayed over land rights disputes with the *quilombo* community (descendents of runaway slaves) near Alcântara.[162] Other collaboration agreements were signed with Argentina, Canada, China, Germany, India, and Israel on projects ranging from night-vision radar (with Germany) to satellite construction (China and Israel). The most expansive collaboration agreement has been with Russia, which will help to create a new generation of Brazilian launch vehicles capable of carrying larger satellites, as well as a liquid-propellant version of the VLS.

Up to 22 launches by 2014 were originally envisioned, which would have made the Alcântara Center one of the world's leading space ports.[163] But following the Alcântara accident, the Brazilian Space Agency decided to develop a completely new series of launch vehicles in collaboration with the Russian Federal Space Agency (RKA). Initially, the project was a private venture led by Russian investors, called the Projecto de Sistemas de Lançamentos Espaciais Orion (Space Launch System Project Orion). The proposed launcher would have had the capacity to lift 6,000 kilograms to GEO or 14,000 kilograms to LEO. The Orion project ultimately folded, but the same Russian investors became part of Brazil's newest launch vehicle project. In November 2004 a memorandum of understanding was signed between Brazil and the RKA to create the Cruzeiro do Sul Program ("Southern Cross," in reference to the constellation on Brazil's flag), which will ultimately produce five different launch vehicles, ranging from light- to heavy-lift capabilities, and will be based on the RD-191 engine developed for the Russian *Angara*. For Brazil's part, it is noteworthy that the joint program includes the collaboration of the Comando-Geral de Tecnologia Aeroespacial (General Command for Aerospace Technology), the Brazilian Air Force's research center for space flight and aviation, which speaks to the continued interest of the Brazilian military in space applications. The program is slated to be operational by 2022.[164] Russia has also agreed to improve telemetry and tracking systems as well as the ground infrastructure at the Alcântara launch facility. Brazil has dedicated US$1 billion for the construction of five additional launch pads capable of carrying out up to 12 foreign commercial launches annually.[165]

The first light-lift rocket of this ambitious program, *VLS Alfa* (an upgraded version of the VLS-1), is scheduled to be flight-tested in 2012 and will be capable of placing a 250-kilogram payload into a 750-kilometer equatorial orbit. It is slated to carry the first completely Brazilian-made satellite in 2013.[166] By the time the last of the five launch vehicles enters service (*Epsilon*, capable of putting four metric tons into orbit), Brazil's Alcântara Launch Center would be one of the world's leading spaceports, and commercial launch revenues would provide a healthy budgetary supplement of as much as US$100 million annually. In carrying out this program, Brazil is positioning itself as an important future player in the satellite launch business, potentially a competitor against both NASA and the ESA. Sérgio Gaudenzi, a former president of the AEB, asserted that despite the program's extraordinarily high costs, this newest chapter in Brazil's space policy is crucial to the country's strategic policies.[167]

In 2008 the Brazilian National Defense Strategy (NDS) paper was released. It contained three overarching goals, all of which will benefit from the expanding space program: (1) reorganization of the armed forces and a restructuring the Brazilian defense industry; (2) continued economic development for national security; (3) revision of the policies governing the armed forces. The NDS explicitly calls for more resources for the country's space program. Brazil not only wishes to develop greater launch capacity, but also wants the space program to propel its broader geopolitical agenda of advancing the country's role on the international stage. The NDS also addresses need for further development of nuclear power, the creation of a nuclear navy, and collaboration with erstwhile rival Argentina to accelerate nuclear fuel processing.[168] Each depends in part on the space program's success, as an ever broader array of satellites will facilitate each objective. But perhaps the most salient rewards of Brazil's ambitious space policy goals are not found in the vacuum of space.

First, becoming a first tier space actor will give added weight to Brazil's ongoing campaign to gain a permanent seat on the United Nations Security Council (UNSC). As one of the so-called G4 countries vying for a seat on the UNSC (along with Germany, India, and Japan), Brazil is positioning itself by holding up its burgeoning space program as additional evidence of its role as a regional power, much as the current UNSC members utilized their unique positions as the victors of World War II and as declared nuclear powers. Since Brazil had formerly renounced the development of nuclear weapons through its ratification in 1998 of the Nuclear Non-Proliferation and Comprehensive Test Ban Treaties (NPT), such a trump card is not currently a viable option, though it was hinted at by President da Silva in 2008.[169]

Brazil's accession to the NPT still allows for the development of nuclear energy. The Resende II uranium enrichment facility, 120 kilometers west of Rio de Janeiro, was opened in 2008. This facility closes the nuclear energy loop for the country, eliminating its dependence on outside sources for enrichment and potentially enabling Brazil to process enough enriched uranium for up to 60 nuclear warheads. A third nuclear power plant, Angra III, is scheduled to be operational by 2015. The Brazilian navy has thrown its hat into the nuclear ring by announcing its intention to construct, with French assistance, a nuclear submarine fleet by 2020. Taken together with the country's more independent space program and growing launch capacity, Brazil is positioning itself well to make a unique argument for inclusion as a permanent member of the UNSC.

Second, an equally important consideration is that the space program, along with nuclear enrichment, will give Brazil ever greater autonomy from US influence, which Brazil has chafed at since the end of World War II.[170] These milestones mark some liberation from previous technological dependence from the United States. Once all the pieces of the space program are in place, Brazil may capture up to 10 percent of world satellite launches in the coming decade, including some of the 40 percent that currently are launched from the United States.

Lastly, a successful space program gives Brazil an additional economic advantage over its Latin American neighbors as well as most other developing states. Brazil already leads Latin America in the number and capacity of its telecommunications and remote-sensing satellites. But a principle obstacle faced by the Brazilian government is not so much technical as bureaucratic. Because of the country's antiquated, protectionist tax structure, important Brazilian satellite producers actively court foreign launchers, thus defeating the country's hard-sought goal of autonomy.[171] Moreover, the country's space technology workforce is small and aging, and not enough Brazilian youth are studying engineering and science. To address this demographic hurdle, the Brazilian government initiated a program in 2011 called "Brazil Science Without Borders," which will provide 75,000 science and technology scholarships to undergraduate and graduate students for study in the United States, Britain, or Germany.

Despite some challenges, and demonstrating a growing and an ever more promising launch ability as well as a home-grown space technology program, Brazil has largely succeeded in creating a young, functioning space program that rivals or surpasses all other developing states with the exceptions of its fellow first tier EMSAs, India and China. Coupled with its broad-based technological and resource bases, Brazil is poised to use this newly acquired launch capacity to further its claim to be a burgeoning world power for the twenty-first century as well as one of the world's leading space powers.

4 Second tier space actors

The only way of discovering the limits of the possible is to venture a little way past them into the impossible.

Arthur C. Clarke[1]

Leaving the quickly evolving and increasingly complex group of first tier space actors, we find an even larger cadre of nation-states that have become convinced of the utility and benefits of investing in space-related activities. The motivations and capabilities of these states run the gamut from a few, such as Iran, that are already perfecting an independent launch ability, to a large number of states for whom the construction and/ or ownership of satellites and their various space applications have become integral components of national security and development strategies.

The countries of the second tier of space actors represent a large range of political, economic, and social systems, but their commonality is that each chose to invest in space applications that have specific direct as well as indirect benefits to their respective national security and developmental goals. But, until South Africa recently broke from its apartheid-era security-oriented launcher programs by announcing the development of a weather satellite system, all the second tier space actors shared a focused space policy that outlined their limited use of space through the development of rudimentary launch systems and basic satellites oriented toward remote sensing, communications, and scientific observation, especially meteorology. Moreover, true to the historical precedents of the use of space set by the DVSAs and the first tier EMSAs, the second tier EMSAs have employed their space programs as a dual-use setup to further their respective ballistic missile and, in a few cases, actual or would-be nuclear weapons programs. The difference between the technologies required to put a satellite into orbit and what is necessary to place a conventional or nuclear warhead on a target hundreds or thousands of kilometers distant are very small indeed, with only minor variations required in guidance systems.

While, in almost every case, these second tier space programs' projects have included plans with dedicated socioeconomic designs, such as remote-sensing satellites to improve agricultural production, the underlying rationale for second tier programs conforms to the central thesis of this book. Given sufficient investment potential, the space activities of most second tier EMSAs began as security-oriented programs that, as was the case with space programs in the developed world, were at best projects in which a country's military has had great institutional interest and occasional participation, and at worst mere window dressing for furthering ballistic missile development programs. However, for some second tier space actors, their bellicose beginnings have transformed over time into purely civilian programs with the sole purpose of contributing to national socioeconomic development.

The Middle East and Africa

Four second tier states of the Middle East and Africa—Iran, Iraq, Israel, and South Africa (see Table 4.1)—show similar trajectories in the development of their space programs, though with some important differences in their implementations. Though separated by geography, government type, and a multitude of other factors, each of these second tier space actors has shared the ambition to utilize space as a "force multiplier," a factor that would dramatically increase the effectiveness of existent security and developmental policies in a way that that terrestrial options alone could not.

Though standing in strident political opposition to each other today, Iran and Israel are the most advanced second tier EMSAs, and both are manifest examples of EMSAs which have chosen to focus their current and potential space power abilities almost exclusively in a bid to ensure their respective national security in less than hospitable geopolitical circumstances. The rapid growth of their respective space programs stands in sharp relief to their relatively recent entry into space-related endeavors. These budding space actors have not yet pursued proportionally much in the way of broader socioeconomic development applications of space technologies, despite public rhetoric to the contrary. So while the Iranian

Table 4.1 Second tier space actors

Country	Space agency	Budget in US$ (year)	Current space activities
Iran	Iranian Space Agency	500 million (2004)	Launcher and satellite construction
Iraq	none currently	none currently	Launcher and satellite construction (previously)
Israel	Israeli Space Agency	50 million (1983); over 100 million with military activities	Launcher and satellite construction
South Africa	South African Space Agency	13 million (2011)	Satellite construction (previously launchers)

government insists that its space program is benign and for developmental and defense purposes only, the history and patterns of its space power development examined thus far strongly suggest that Iran instead shrouds its efforts to improve ballistic missiles within the framework of a civil space program.[2] Israel, on the other hand, simply obfuscates all its space efforts under a veil of almost complete non-disclosure, especially where its space program intersects with its nuclear weapons program. Both countries' programs in missile and space technology have achieved an impressive level of sophistication considering their relative youth and comparatively modest budgets, and show every sign of accelerating and growing in the coming years in size and complexity.

On the other hand, though formerly on a similar trajectory of development for their space programs, Iraq and South Africa offer telling examples of second tier space actors that sought and acquired roughly the same space capabilities but which have now eschewed and/or lost some or all of their space capabilities; South Africa has only recent revived its space program, under completely revised civilian leadership with an exclusively developmental mission.

Iran's revolutionary orthodoxy

Much like more traditional space actors who have benefited militarily from advancements in their space programs, Iran has also used the dual-use nature of its space program as a façade to conceal the development of its conventional missile systems, simultaneously demonstrating steady progress in the range, power, and sophistication of the country's launchers. But the *de rigueur* debates surrounding the ambitions underlying Iran's ballistic missile and space programs miss the point, since these programs are not modern at all, but predate the theocratic regime that has ruled the country since 1979.

Iran's attempts to utilize space originated with the government of Shah Mohammad Reza Pahlavi, who ruled from 1941 to 1979. Though the missile-nuclear-space (MNS) triad was not an official facet of the Shah's White Revolution (which focused on reform programs to legitimize his rule), the MNS triad programs were concurrent and were intended to raise Iran's image and standing as a modern nation-state. Early on, the Shah's missile and space aspirations led Iran to become one of the founding 24 members in 1959 of the United Nations' Committee on the Peaceful Uses of Outer Space (UNCOPUOS) and to sign the Outer Space Treaty in 1967. In 1969 Iran founded the country's national Telecommunication Manufacturing Company, with a 40 percent investment by Germany's Siemens AG.[3] Iran was also an original member of the International Telecommunications Satellite Organization (Intelsat) in 1970, became a member of the International Telecommunications Union in 1971, and was an early partner with United States in processing remote-sensing imagery from US Landsat satellites.

Under the Shah, Iran took its first tentative steps toward planning a state communications satellite system through the founding of the Iran Telecommunication Research Center in 1970 and the Telecommunications Company of Iran in 1971. These organizations were involved in research that was to have eventually led to an Iranian reconnaissance satellite.[4] Reflecting Iran's early multi-purpose designs for its space program, the Office for Collecting Satellite Data was set up in 1974 within Iran's Management and Planning Organization, which established the use of remote-sensing technology. It was renamed the Iranian Remote Sensing Center later the same year and established cooperative arrangements to process US Landsat images. Iran opened a data-receiving station in Mahdasht in 1978 with the initial help of the US firm General Electric, though GE left before the station's completion because of the looming revolution (the Mahdasht facility was subsequently renovated and modernized in 2003).[5]

Of equal importance to the birth of Iran's modern missile and space programs were the country's historically poor relations with its neighbors during the Cold War. Out of fear that pro-Soviet Arab states might eventually attack Western-leaning Iran, especially Egypt under its nationalist leader Gamal Abdel Nasser, Iran spent record amounts on weaponry, concerned that the US might not fully intervene in a "regional conflict." Ironically, Iran purchased over US$8 billion (US$31 billion in 2011 dollars) worth of weaponry from the US between 1973–76.[6] Exact figures were not published at the time, but the Stockholm International Peace Research Institute estimated that, for example, Iran spent US$14.6 billion in 1976 on weapon systems and perhaps as much as US$31billion throughout the mid-1970s, representing between 19 to 30 percent of the country's total annual budget.[7] Of particular concern to Iran was its neighbor, Iraq, which had become a major purchaser of Soviet Scud missiles. To counter the potential of Iraq's newest acquisitions, the Shah's Iran first sought to acquire US Pershing missiles and then to develop a nuclear deterrent.

Like India, Iran pursued a nuclear program that had been jump-started via the United States' "Atoms for Peace" program. A US-supplied five-megawatt nuclear research reactor was built and became operational in 1967, and a contract was signed in 1975 with West Germany's Kraftwerk Union AG to build Iran's first full-scale nuclear reactor in what was to have been part of an eventual network of 20 reactors. In 1976, the Ford administration offered Iran the opportunity to purchase US reprocessing technology that would have permitted the extraction of plutonium from the nuclear fuel, thus completing the nuclear cycle and beginning a nuclear weapons program for Iran.[8] With this seed technology in place, foundational research into other nuclear applications began and Iran entered into an oil-for-technology agreement with South Africa for nuclear fuel enrichment technology.[9] After being rebuffed by the US in its bid to purchase Pershing missiles because of US fears that the missiles would carry future Iranian nuclear warheads, Iran turned—in a move whose irony is now apparent—to Israel for assistance.

Israeli-Iranian cooperation was not always antithetical. The beginnings of a marriage of convenience between the two states had already appeared in the 1950s out of the changing geopolitical circumstances, as more Arab states becoming Soviet client states, and given Israel's need for Iranian oil. In April 1977 Iran signed a US$1-billion oil-for-technology agreement with Israel called "Project Flower," which would provide Israeli assistance and know-how to build Iranian surface-to-surface, ballistic, and even submarine-launched missiles.[10] Israel also agreed to build the necessary missile production facilities in Iran, which would have allowed Israel to surreptitiously circumvent US technology restrictions by using Iranian missiles for third-party sales.[11] Construction began on a missile assembly facility near Sirjan in south-central Iran and a missile test range in nearby Rafsanjan. In exchange, Iran made a US$280-million down-payment in petroleum. The missiles were designed to be capable of carrying nuclear warheads, but the 1979 Islamic Revolution put an end to this unlikely cooperation.

However, the ousting of the Shah did not diminish the Iranian government's perception of the need for missile technology, and in fact, the origins of the country's current space program are largely found in the change of government. Though initially resistant to the idea of missile investment because of the funds needed for promised social reforms, the new Iranian revolutionary government began work on a strategic deterrent to forestall any possible US attempt to reinstall the Shah. The new government benefited from continued Israeli military sales, which were prolonged by Israel in the hope that they would curry favor with the new regime.[12] But the sustaining catalyst for the Islamic Republic's accelerated development of the missile-nuclear-space triad was the ruinous Iran–Iraq War of 1980–88.

During this costly war for regional hegemony, Iranian cities were pummeled by over 500 Iraqi Scud and Al-Hussein missile attacks, which forced an abrupt reevaluation of spending priorities by the revolutionary government and prompted them to redirect spending toward military needs. Between 1980 and 1986, military expenditures rose from 6.6 to 7.2 percent of GDP and constituted almost 30 percent of central government expenditures.[13] The missile program to counter the Iraqi advantage was a primary recipient of the military budget increase. By 1986, with purchases of Scud missiles from Libya, Syria, and later North Korea, and the building of the Hwasŏng-5 missile assembly plant with North Korean help, Iran entered the previously one-side missile volleys with Iraq. Toward the end of the war, in the so-called War of the Cities, over 170 Iranian missiles rained down on Baghdad. Among these missiles were 18 Scuds, which would later become the launcher foundation of the Iranian space program.[14] Ending in a draw, the war gave Iran every reason to become self-reliant in missile technology. Post-bellum, Iran continued the improvement of its missile program with ongoing technology transfers from China, Russia, and North Korea in the areas of guidance, propulsion, and training.[15]

Iran's space ambitions began to gain momentum in the mid-1990s, through the expansion of cooperative programs with more established space actors and decisions to institutionalize the production of space technology within Iran itself. An agreement with India to share remote-sensing imagery, an agreement with Russia to build a civilian communications satellite (to be launched by Russia), and a mutual defense treaty with India are but a few examples of Iran's ventures during the decade.[16] On its own, Iran also made plans for a national satellite communications system to be put in geosynchronous orbit. This plan was publicly announced in 1999 and bids were solicited from India, Russia, and France for the *Zohreh* (the planet Venus in Farsi) project, which had its beginnings in the Shah's unfulfilled plans for a four-satellite telecommunications system in the 1970s. In the meantime, Iran leased bandwidth from the Russian *Gorizont* satellite system to provide communications to the whole of the country, especially Iran's more remote mountainous regions. Finally in 1998, two key developments helped to push Iran into space.

First, an intra-agency agreement for US$10 million was entered into in April 2003 by the Ministry of Higher Education and the Ministry of Post, Telegraph, and Telephone with the Italian company Carlo Gavazzi Space for the design, construction, and eventual launch of an Earth observation satellite called *Mesbah* (Lantern).[17] However, *Mesbah* was put on indefinite hold because Italy has been pressured not to deliver the finished satellite by UN Security Council resolutions against Iran. Second, a memorandum of understanding was signed by Iran with China, South Korea, Mongolia, Pakistan, and Thailand, which outlined a joint venture to build two small multi-mission satellites (SMMS). The SMMSs were intended give Iran a semi-autonomous space-imaging capability. This first goal, though unfulfilled, was an important step toward the institutionalization of the modern Iranian space program.

The beginnings of Iran's autonomous launch capability were the original Scud missiles acquired and developed during the Iran–Iraq War. Renamed *Shahab* ("Meteor"), the missile was a variant of the Scud-C and was introduced in the late 1980s. It was the first in a series of *Shahab* missiles to come. The follow-on *Shahab-2* was unveiled in 1990 and, by the late 1990s, Iran had acquired the necessary materials, technology, and expertise to begin production of its first medium-range missile, the *Shahab-3*, which had its first test flight in 1998—the same year that Iran announced its new space program.

The *Shahab-3* reportedly benefited not only from North Korean, Russian, and Chinese assistance but also from a North Korean *No Dong* engine that gave the missile a range of over 1,200 kilometers, thus making it Iran's first true space-capable launcher.[18] Though a subsequent flight test in 2000 ended in failure, the program advanced haltingly, despite repeated attempts by the United States to pressure Pyongyang to end its missile assistance to Tehran. Undeterred, North Korean added expertise to indigenous Iranian modifications, which together with the ever-expanding domestic technology industry led to the creation of an improved *Shahab-3*, which was test-fired in late 2004. Its launch served the dual purpose of testing the missile and broadcasting the message of Iran's newly gained prowess in weapons systems to the region and to the United States. The newer and more sophisticated *Shahab-4*, based on the 1950s Soviet R-12 MRBM (the same type based in Cuba during the 1962 Missile Crisis), has been publicly characterized by Iran as its newest space launch vehicle, though it probably also serves as the basis for intermediate-range ballistic missiles.

Believing Iran to be the next target of the United States after the 2003 American invasion of Iraq, Iranian Defense Minister Ali Shamkhani declared that Iran would "penetrate the stratosphere" to ensure that "the region cannot be used against us by any outside force."[19] His warning punctuated Iran's accelerated program to join the space actors with auto-nomous launch capability. To provide legitimacy to the outside world as well as the coordination of Iran's domestic space industries, the Iranian Space Agency (ISA) was established in the same year. The ISA's stated offi-cial mission is to serve as a coordinating entity for space policy initiatives, research on space technology, remote-sensing projects, and participation in space exploration.[20] The titular head of the ISA was designated as the president of Iran, though the missile program upon which the space program is based remains under the command of the Iranian Revolution-ary Guard.

In a situation unique to Iran's political structure, the ISA technically exists as an organization under the Supreme Aerospace Council (SAC), which is comprised of ministers of defense, foreign affairs, and communi-cations, among others. The ISA's president is the secretary of the SAC, though under a 2008 amendment Iran's president no longer heads the ISA; instead it is supposed to be administered by the Ministry of Commu-nication and Information Technology, which is supposed to reduce its autonomy but submit its activities to greater oversight.[21] One of the responsibilities of the Supreme Aerospace Council is to manage the coun-try's various space-related programs and to promote partnerships with foreign contractors.[22] In 2004, with a budget allocation of US$500 million, the ISA ardently declared that Iran would become a major space actor within a decade.[23] Iran announced its intention of launching the Islamic Republic's first indigenous satellite and the new 2,000-kilometer-range *Shahab-5* launcher was successfully tested.

Of equal importance, the founding of the ISA signaled a significant overall policy redirection. Concurrently with the development of the coun-try's first space launcher, Iranian President Mahmoud Ahmadinejad announced in 2006 that Iran had joined the group of countries capable of uranium enrichment, though he was quick to add that the program was only for civilian energy production.[24] Western intelligence sources antici-pate the introduction of an Iranian ICBM before 2015, and postulate that Iran's space program is merely cover to allow testing of these missiles.[25] This policy approach is central to the Iranian deterrent posture because of Israel's undeclared but universally recognized nuclear weapons capability. Iran is also keen to acquire sufficient missile capability to counter hostile American foreign policy towards it. Ahmadinejad has pointedly declared that Islam should eradicate what he terms "300 years of Western imperial-ism," and he clearly envisions Iran at the forefront of such a rollback.

In line with these predictions, in the first decade of the twenty-first century Iran made great strides to establish an autonomous space program, using the Shahab missile as the foundation of its launcher program. Iran shows every sign of having effectively mastered the missile technology necessary to continue its space program without further signifi-cant outside help in the near term.[26] Iran's entry into space is recent, but is firmly entrenched in the progression through the aforementioned first tier criteria. Only the development of a nuclear weapons program is lacking to complete the MNS triad so frequently followed by other major space actors, and by all indications this missing link is forthcoming.

The first product to emerge from Iran's revamped space program was the country's first indigenously produced satellite, a very low-resolution (250 m resolution) reconnaissance satellite called *Sinah-1*. The original plan envisioned the use of an indigenous launcher based on the *Shahab-3*, but technical snags forced Iran to turn to Russia. For US$8million, the *Sinah-1* was sent aloft in October 2005 aboard a Russian *Proton* launcher and put into a sun-synchronous near-polar orbit, which would bring the satellite over Iran and the surrounding region once per orbit. The satel-lite's launch made Iran the forty-third state to own a satellite. While Iran publicly announced its intention to use the satellite to provide imagery of Iran and to monitor natural disasters, there are reports that strongly suggest its covert military mission, since imagery was not slated to be shared with civilian agencies.[27] Russia and Iran thereafter signed another deal, this time to revive the more than 30-year-old *Zohreh* project, now transformed as a US$132-million GEO telecommunications satellite system. The work is still pending, with a tentative launch date set for 2014.

To fill in the missing launcher link in its space program, on 17 August 2008 Iran tested its indigenous *Safir* (Messenger) launcher (originally named *Kavoshgar*), a two-stage rocket roughly based on the North Korean *Taepodong-1* and the product of almost two decades of cooperation between the two countries.[28] State-run media reported that the low-orbit research rocket was capable of carrying a satellite, though non-Iranian observers note that it never reached its intended orbit. The first stage of the *Safir* was nearly indistinguishable from the *Shahab-3* ballistic missile, reflecting the launcher's military origins.[29] Carrying a dummy satellite, the launch coincided with the inauguration of the country's new space center, the Chinese-built Missile and Space Center in Semnan, 200 kilometers east of Tehran. The Semnan facility possesses a broad array of domestic factories, missile research centers, and test centers. Other launch sites in Iran include Emamshahr in northeastern Iran, Qom in western Iran, and a brand-new site near Semnan, built with North Korean assistance. To deepen the institutionalization of its space program, Iranian engineering students have been attending the Moscow Aviation Institute, Russia's leading missile school and the one that had trained Chinese rocket engineers just a generation earlier.[30] Russia has also supplied Iran with nuclear reactors, rocket engines, training, and test equipment, such as wind tunnels. In addition, Iran has North-Korean-built liquid fuel plants at Esfahan and Sirjan.

After the tentative but important first step of building *Safir*, Iran finally joined the growing group of countries with autonomous launch capability on 2 February 2009 by launching a domestically produced, albeit simple, 27-kilogram satellite called *Omid* (Hope) atop its newer, longer-ranged *Safir-2* launcher. In his announcement of the launch to the country, Iranian president Mahmoud Ahmadinejad declared that "your children have put the first indigenous satellite into orbit … with this launch the Islamic Republic of Iran has officially achieved a presence in space."[31] In addition, Iran offered help to any Muslim country that wanted to establish its own space program. Ahmadinejad's statement, the offer of assistance, and the timing of *Omid*'s launch during the 10-day celebration of the thirtieth anniversary of the Iranian Revolution reveal the intense political calculation associated with the project, especially given that Ahmadinejad's election platform promised more modernization and economic benefits for the average Iranian. Though *Omid* was only equipped with basic radio transmitters, this first independent step into space makes Iran the eighth country to independently launch its own satellite into orbit (see Table 3.1). The launch of *Omid* was also the springboard for the announcement of the country's 12-year space plan, which includes proposals for the country's first manned orbital flight between 2017 and 2021 and an Iranian astronaut on the moon by 2025.

Even though most analysts dismiss an Iranian lunar program as merely political rhetoric, the public dimension of the program accomplishes the same domestic and international prestige-building goals as did the Apollo and Sputnik programs. For these reasons, the feat has been described and assailed by Western critics as Iran's *Sputnik*, a disparagement that portrays the achievement as a Trojan horse for the country's ballistic missile program. Nonetheless, the success of *Omid* demonstrates that Iranian space scientists have mastered the separation of payload from missile in space and the placement of a payload into the right orbit. This achievement is doubly significant since any multi-stage space launcher is in effect a potential ICBM, capable of reaching the limit of its range with little variation in its guidance systems. More importantly, like *Sputnik* before it, *Omid* served as a demonstration of growing Iranian technological competency and signaled Iran's expected greater political leverage in the region. Lastly, the launch was clearly meant to help the matter of prestige-building among the Iranian people.

Along with the overriding security imperatives associated with the Iranian space program, a fair number of socioeconomic objectives have been addressed as well. Through the Iranian Remote Sensing Center (originally established under the Shah in 1973 to utilize Landsat data), ground imagery is shared and coordinated in the study of mineralogy, forestry, oceanography, cartography, and especially geology, important given the region's seismic propensities.[32] In addition, since Iran is the world's fifth-largest exporter of petroleum, it not surprising that the space program should be utilized in the service of the development and utilization of remote imaging for the petroleum sector. The other principal agency related to the development of the space program is the Aerospace Research Institute (ARI), which produces aerodynamic designs and does analysis of launch vehicles. In addition, at least seven Iranian universities now offer aerospace-related programs in remote sensing and geographic information systems.[33]

Iran has taken great pains to portray its space program in the best possible light to the outside world. In 2008, the UN Committee on the Peaceful Uses of Outer Space (COPUOS) elected Ahmad Talebzadeh, the president of the Iranian Space Agency, as the chairman of the Legal Subcommittee for the period of 2010–11. Given the committee's stated mission to ensure equal, peaceful access to space for all states, the election was undoubtedly a public relations bonanza for selling the Iranian space program. The election coincided with Iran's unveiling in 2008 of its now fully integrated space program and space infrastructure.

Iran's second indigenously built satellite, *Rasad-1* (Observation-1), was launched on 17 June 2011 atop the new and improved *Safir-B1* launcher. The satellite was identified by the Iranian leadership as the country's first spy satellite.[34] Iran plans to launch three additional satellites in the following two years. These satellites will include *Mesbah-2*, a telecommunications and navigation satellite, and *Mehr Navid-e Elm-o Sanat*, a solar-powered telecommunications satellite. In addition, many reports indicate the development of a newer, more powerful launcher under development, *Shahab-5*, which may include both North Korean and Russian technology to increase its payload capacity to 1,000 kilograms and its range to over 4,300 kilometers.[35]

Iran's space ambitions now receive a steady funding stream. Beginning in 2005, as part of the country's Five-Year Development Plan, the space program was allocated US$422 million. To eventually accomplish the goal of manned flight, the new *Simorgh* launch vehicle was unveiled in 2010. At

27 meters high and 85 tons, it carries a liquid fuel propulsion system with the lift capacity to carry a 100-kilogram satellite some 500 kilometers into orbit. As a first step toward sending humans into space, Iran announced in 2010 that it had successfully sent into orbit a rat, two turtles, and worms, mirroring US and Soviet test flights of the 1950s.[36]

Critics have charged that Iran's rapidly expanding space program is merely a façade for the surreptitious improvement of its ICBMs to eventually carry nuclear weapons, criticisms that Tehran has denied. Indeed, the Iranian Ministry of Defense's attempt during the mid-2000s to develop its military reconnaissance satellite called *Sepehr* to provide early warning capability—presumably from Israel—strongly reinforces the argument that Iran is eager to exploit the benefits of its growing space capabilities for national security purposes.[37] Probably in an effort to defuse these charges, Iran has also begun to engage in non-military oriented space endeavors such as a joint project with China and Thailand called *Environment 1*, launched by a Chinese *Long March 2C* on 6 September 2008. Its mission is to jointly track natural disasters in the respective countries and to promote and develop the space programs of other Asian countries.

Nonetheless, it stands to reason—given the dual-use nature of space launchers—that obfuscation is precisely what is happening, since this precedent was set over 50 years by the original space actors. Iranian officials, in fact, are quite open that the Shahab ballistic missile is the basis for Iran's satellite launch capability.[38] Unless Western diplomatic or military preventative measures prevail, Iran's nuclear program will undoubtedly continue and Iran will eventually achieve a full MSN triad.

Moreover, the Iranian space program plays a very important domestic political role. The mission and lofty goals of the IRA are not only to provide technological advancement for the missile and space programs, but just as important, to assure the position of the mullahs, as the political situation is, by some estimates, as volatile now as it was before the revolution.[39] The mullahs are under increasing criticism and scrutiny from a new generation of young people who question the authoritarian nature of the regime. Therefore, a fully constituted MNS triad will not only serve Iran's military needs but will possibly safeguard the position of the regime through prestige-building and enhanced national pride.

The speed with which Iran has pursued its missile and space programs coincides with its equally bold endeavors in other areas of national strategic interest. For example, Iran is one of the few countries in the world that now possesses the new, potentially devastating, supercavitating torpedo called the Hoot (Whale). Reverse-engineered from the Russian VA-111 Shkval torpedo, which uses expended gas to reduce water friction, the Hoot torpedo is capable of reaching underwater speeds of 360 kph or more.[40] Much as the 1862 US Civil War naval battle between the ironclads CSS Virginia and USS Monitor made the world's wooden navies obsolete overnight, these supercavitating torpedoes could potentially do the same to the surface ships of its adversaries. Iran's short-term goal is to deter the United States and gain more freedom of action to become a nuclear power. Its longer-term goal is to project its influence beyond its borders, using its space program as a primary vehicle. Iran's progress so far clearly demonstrates that Iran stands at the crest of competing the MSN development cycle, thus following the path blazed before it by the established space powers, and perhaps entering into first tier status in the coming generation.

Israel: Samson on high

While not currently a developing country in terms of GDP per capita, Israel's rise as a space actor has paralleled the aforementioned *raison d'être* of EMSA space programs and, more importantly, has been strongly motivated, affected, and even intertwined with the growth of other EMSAs. Therefore, the story of the Israeli space program is an essential part of an inclusive understanding of the evolution of space policy throughout the developing world, but with special import in the Middle East and Africa. As previously noted with Iran, a great deal of the current motivation for that country's current missile and space programs is attributed to its persistent tension, if not outright animosity, toward Israel following the 1979 Iranian Revolution.

As would be anticipated under the rational-actor model, Israel is undoubtedly preparing to act in what it believes is its national interest given its geopolitical isolation in the region. Having fought four major wars with its Arab neighbors since declaring independence in 1948, Israel has been unavoidably forward-thinking in its foreign policy, particularly in defense and, most recently, in its space policy. The most important point of departure in understanding Israel's drive toward space stems from the October 1973 Yom Kippur War, when Egypt and Syria launched a surprise attack against Israel on the most holy of Jewish holidays.

When these combined Arab armies launched their invasion, Israel initially found itself dreadfully unprepared (despite an unheeded warning from Jordan). With most of its regular army away on holiday and facing a combined onslaught of over 3,000 tanks, 2,000 artillery pieces, and one million Arab soldiers, Israel confronted the most dire threat yet in its short existence as a sovereign state. Despite Israel's desperate pleas for immediate US weapon shipments, the US was slow to respond because US Secretary of State Henry Kissinger underestimated the situation. The crisis brought Israeli Prime Minister Golda Meir to the point of ordering the assembly of 13 of the country's 20-kiloton tactical nuclear devices and the preparation of Israel's Jericho ballistic missiles for nuclear retaliation (coded as the "end of the third temple"), if the situation deteriorated further and total defeat seemed eminent.[41] Though over 10,000 metric tons of US shipments did ultimately arrive and Israel successfully fended off the invasion, the resulting peace accords with Egypt denied Israel its previous access to important intelligence resources in the Sinai Peninsula.

The United States' delay in providing vital matériel, weapons, and intelligence (including overflights by the SR-71 Blackbird) until three days after the invasion began convinced Israeli leaders that such utter dependence on the US for future supplies, technology, and intelligence was no longer prudent. Thus, the impetus for the expansion of Israel's home technology industry and, ultimately, its space program were in direct response to its perceived national interest to limit its outside dependence and to guarantee its sovereignty. With this new imperative, and given Israel's small size and limited resources, the development of new indigenous sources of intelligence, communications, and military technology was given the utmost priority. These needs became even more acute as the United States, long Israel's most loyal strategic partner, decided to be more selective in what satellite imagery it would share with Israel.[42]

The response from Israel was one of quiet indignation. Meir Amit, a former chief of Mossad (Israel's intelligence agency), complained that "if you are fed from the crumbs of others according to their whim, this is very inconvenient and very difficult," and resentfully equated the situation with a patron-client relationship.[43] The United States' decision to begin to provide economic and military aid to Egypt was the last straw in prompting Israel's move to diversify its strategic partnerships and policies. Between 1973 and 1981, the Israeli arms industry expanded almost fifteen-fold, encompassing practically every area of technology application, especially in the improvement of missile technology for both nuclear ballistic missiles and space launchers. However, the kernel of the Israeli space program is found more than a decade earlier in the establishment in 1960 of the National Committee for Space Research (NCSR), a group dedicated to the research and development of space-related sciences, and to demonstrating Israeli capabilities to its antagonistic neighbors at the time, especially Egypt.

In 1961, Israel launched the solid-fuel, two-stage sounding rocket *Shavit* on a meteorological mission, and an agreement for further missile development was signed with the French aerospace conglomerate Dassault, which provided crucial technology, such as the Dassault MD-620, which Israel would eventually use to refine its *Jericho* intermediate-range ballistic missile (IRBM). When the program was cut short by the 1967 Six-Day War and a subsequent French missile embargo, the US replaced France as Israel's principal foreign source of high technology. Newly supplied, Israel then developed warhead and guidance systems, but it was not until the 1977 political victory of the right-wing Likud party and the leadership of new Prime Minister Menachem Begin that Israel's space ambitions fully materialized. The Begin government approved the establishment of a formal space program in July 1981 and the process of converting the *Jericho-2* IRBM into the *Shavit* space launcher began in earnest.

In November 1982, the Ministry of Science and Technology announced the formation of Israel's official space agency to coordinate and supervise a national space program. The next year, the Israeli Space Agency (ISA) was founded as a division within the Israeli Defense Force (IDF), thus unambiguously paying homage to the space program's legacy and implying its continued security-oriented purpose. This relationship is stressed further by the relatively small annual budget of the ISA, estimated at US$50–60 million—much too paltry on its own to support launcher and satellite programs without an infusion of funds from the IDF.[44] This orientation is further clarified by the ISA's mission, which states space research and exploration is an essential instrument for the defense of life on Earth; the lever for technological progress; the key to existing in a modern society; essential for developing an economy based on knowledge; and the central attraction for scientific and qualified human resources.[45] The ISA has sought to benefit from working relations with a multitude of space agencies from both the developed and developing worlds. Space technology cooperation agreements have been signed with France, Canada, India, Germany, Ukraine, Russia, and Netherlands, and agreements are pending approval with Chile, Brazil, and South Korea.[46]

Reflecting the trend among both developed space actors and other second tier EMSAs, Israel has made extensive use of private contractors to fulfill its aerospace needs. In 1983, in a classic example of government/private sector collaboration, the Israeli military employed a private contractor to develop the necessary framework for the country's first spy satellite. This most utilized contractor is Israel Aerospace Industries (ISI), which designs and manufactures a wide array of technologies, including missiles, radars, planes, ships, and space systems for both military and civilian applications.[47] In 1988, Israel became the eighth state to gain independent launch capability by putting the ISI-manufactured *Ofeq 1* reconnaissance satellite into low Earth orbit aboard Israel's indigenous *Shavit* launcher. Nine more *Ofeq* spy satellites followed from 1990 to 2010. Israel's first ostensibly civilian-oriented remote-sensing platform was launched in 2000 from Russia. The Earth Resources Observation Satellite (EROS-A) has respectable 1.5 meter resolution; advertised to provide high-quality digital photography for sale to anyone, it "symbolizes the inauguration of [Israel's] civilian imaging satellite program."[48] But its *Ofeq*-based design and its construction by the Tel Aviv-based private company ImageSat International, an Israeli military contractor, weakens the civilian-only claim and strongly suggests instead a prescribed dual-use mission. Israel's only astronaut, Ilan Ramon, perished aboard the ill-fated *Columbia* space shuttle on 1 February 2003.

As of June 2010, with the launch of *Ofeq 9*, Israel maintains at least six spy satellites in orbit. This fits the mission of the ISA, as the lion's share of Israel's space efforts have been solidly in line with the country's perceived security needs vis-à-vis its neighbors. Only two of Israel's more than

18 satellite launch attempts have been for commercial purposes. Most have been reconnaissance satellites, whose orbits have been specifically calculated to pass over the country's neighbors, especially Iran. Israel's impressive though imperfect launch capability is a direct consequence of the country's almost maniacal, though historically necessary, obsession with national security. Israel's increasing competence in space systems has created a regional space mini-drama of sorts, forming the basis for diplomatic wrangling between Israel, Iran, and India over launch rights, national airspace, and national sovereignty, predictable given the very limited launch trajectories afforded Israel by its geography.[49]

Despite the overwhelming security focus of the space program, the ISA and the ISI have increasingly engaged in some purely scientific, socioeconomic, and commercial applications of Israel's space technologies, though these have been launched by other space agencies. Besides the EROS program, examples include the *Amos* series of communication satellites launched in 1996 (launched by ESA), 2003, and 2008 into geosynchronous orbit to provide service to customers in Eastern Europe and the Middle East, and *Techsat II*, a microsatellite launched in 1998. Weighing only 45 kilograms, *Techsat II* possessed miniature cameras and technology for communications, remote sensing, astronomy, and geosciences, such as topographical projects that use ground-penetrating radar to identify water sources to irrigate the Negev Desert, which covers over half of Israel's territory.[50]

Nonetheless, Israel's state space program plays an integral part in the bigger picture of Israeli security policy and completes its MNS triad. Israel is the world's largest undeclared nuclear weapons state, and the secrecy of the program was facilitated by a tacit understanding between the United States and Israel that the Jewish state would surreptitiously own nuclear devices but not test them. In exchange, the US would do three things: (1) ignore the weapons' existence; (2) exert no pressure on Israel to sign the Nuclear Non-proliferation Treaty (NPT), and (3) end previously secret annual visits by the US to Israel's nuclear facility at Dimona.[51] According to a recently declassified US General Accounting Office document, the first uranium (or plutonium) used for Israel's weapons was clandestinely provided by the United States.[52]

Israel is one of only three nuclear-capable countries not to have signed the NPT (along with India and Pakistan). Though its nuclear capability is neither confirmed nor denied by the Israeli government, under a policy known as "nuclear opacity," information and furtive photos from Israeli whistle-blower Mordechai Vanunu (imprisoned for 18 years by Israel for treason), as well as various foreign intelligence estimates, state that Israel possesses at least 100 and perhaps as many as 250 thermonuclear warheads.[53] This situation, together with the well-understood Israeli nuclear retaliation policy which promises massive retaliation for an attack on Israel (the so-called Samson Option), means that Iran's rationale for the dual development, dual-use space program is clear and profoundly understandable from a balance-of-power perspective. The space program has been developed to assure the existence of the Israeli state. Furthermore, the backbone of Israel's ballistic and space programs, the Shavit launcher, is itself superb illustration of cooperation outside the traditional space actor sphere of collaboration, in forming a long-lasting and mutually beneficial security arrangement with South Africa.

South Africa: apartheid, freedom and space

As with many other space actors, the early foundations of South Africa's experimentation in rocketry and space activities were in the work of amateur enthusiasts. Mirroring pre-World War II Germany, the South African Interplanetary Society was founded in Johannesburg in 1953, and this group's rocket experimentations laid the groundwork for the creation in 1959 of the South African Rocket Research Group (SARRG). Between 1959 and 1963, the SARRG conducted 102 static firing tests and launched 528 rockets, including some with two and five stages.[54] But precisely at the time the group's newest rocket was ready to break the 100-kilometer Kármán line, government inspectors shut down the project, not coincidentally in the same year that the South African military began the development of tactical surface-to-air missiles.

These initial amateur efforts were supplanted by the South African government's increasing cooperation with the United States to provide telemetry services for space probes during the space race. Such was the case with the NASA-built 26-meter dish at Deep Space Network Station 51, completed in 1961 at Hartebeeshoek, near Johannesburg, which was used through the 1960s for telemetry downloads from NASA's Mariner probes to Venus and Mars and various moon probes. But by the early 1960s, outrage in many Western countries against the Afrikaner government's apartheid policies led to the reduction and eventual cancellation of rocket and space science cooperation. Because of this setback and its increasingly isolated position, South Africa promoted the creation of a domestic military-industrial complex. But the key to the country's eventual success in developing its MNS triad came in the form of an improbable partnership between the pariah apartheid regime and the Jewish state of Israel.

Both countries were relatively technologically advanced, had regionally powerful militaries, and were surrounded by antagonistic neighbors. The fact that they stood in diametrical opposition to each other in terms of ideology was not sufficient to hamper their cooperation on matters of national security in the areas of nuclear weapons, missiles, and space launcher development. So while Israel's pro-leftist, anti-colonial policies led the Jewish state to support many other African states because of staunch Israeli opposition to the legacies of colonial discrimination, South Africa's apartheid government continued to support a particularly virulent version of systematic discrimination now well known as apartheid. This *manus manum lavat* arrangement lasted until the early 1990s.[55]

The South African government's world view and policymaking were driven by Prime Minister P.W. Botha's paranoia about Soviet expansionism in southern Africa and by the government's desperation over international revulsion at apartheid. Technical cooperation between South African and Israel increased during the 1970s and the early 1980s, as the Afrikaner government desperately sought the means to withstand the increasing internal and external pressures to dismantle its system of apartheid, and to combat the installation of Soviet-backed governments in the newly independent neighboring states of Angola and Mozambique.[56] All defensive options were considered, and domestic research was pursued in ballistic missiles as well as nuclear, chemical, and even biological weapons.[57] After a series of high-level meetings, South Africa entered into a covert military alliance with Israel in 1975.

An agreement code-named Secment created the alliance between the two states, which stipulated that Israel would provide South Africa with advanced rocketry and nuclear weapons technology, under the project name Chalet.[58] Israel would also serve as an intermediary for third-party technology sales to the increasingly isolated South African regime, and the two states would jointly develop new defense technologies. In exchange, South Africa would provide Israel with the uranium ore and test facilities necessary for the Israeli nuclear weapons program. Though South Africa had produced its own short-range tactical rockets since the 1960s, with Israel's help and the purchase of licensing agreements, South Africa's state-owned arms manufacturer, Armaments Corporation (Armscor), manufactured the RSA (Republic of South Africa) missile series for "communication, commercial, industrial, and military purposes."[59] The RSA was a 23-metric-ton carbon-copy of the *Shavit*, and, on paper, was capable of putting up to 330 kilograms into orbit. Three successful test launches took place at the Overberg Test Range, 200 kilometers east of Cape Town, in 1989 and 1990, with each reaching an apogee of 300 kilometers.

To disguise the missiles' development, South Africa's official R5b space program was founded, which would be the public face for the missile program. Two primary launchers were planned. The RSA-3 launcher began as an intended ballistic missile deterrent against the encroachment of the Soviet Union into southern Africa. The RSA-4 was a planned inheritor to the RSA-3 and was to be a larger, more advanced, four-stage launcher capable of lifting up to 570-kilogram payloads worldwide. Four launchers were eventually fabricated and were set to put South Africa's *Greensat* land resources satellite into orbit in support of commercial planning (the program was eventually cancelled in 1990). Three RSA-3s were test-launched, though in this case, under a 1979 secret executive order from Botha, the space program was at this point officially a front for military experiments to generate a ballistic missile program that would provide increased delivery range of the country's rudimentary nuclear deterrent.[60]

With its beginnings in a classified 1973 cooperative technical arrangement with the West German firm Steinkohlen Elektrizitats AG, the South African nuclear weapons program was the *raison d'être* for the space program, and it was ultimately successful in creating seven basic (gun-type fission) uranium nuclear bombs with yields between 10 to 18 kilotons. The program produced a brief international furor as a result of the so-called "Vela Incident" on 22 September 1979, named for the US observation satellite *Hotel Vela*, which recorded a "double-flash" in the South Indian Ocean, a phenomenon typically indicative of a nuclear weapons test.[61] Numerous sources identify the event as a likely joint South Africa–Israel nuclear test.[62] The South African nuclear program was terminated in 1989 following Cuba's withdrawal of troops from Angola, the fall of the Soviet Union, and the independence of Namibia, though the South African government's public acknowledgement of the program would not come until 2003.[63] At the same time, with the fall of apartheid, US leniency toward South Africa's missile program evaporated, forcing South Africa to sign the Nuclear Non-Proliferation Treaty in 1991 and cancel its cooperation with Israel. As a result, South Africa holds the distinction of being the only country to build nuclear weapons and subsequently give them up voluntarily.

In parallel with its nuclear drawdown, South Africa announced in 1993 the termination of the R5b program, which was to have produced Africa's first indigenous space launch vehicle and, concurrently, the continent's first homegrown ICBM force. The outgoing De Klerk government signed a diplomatic letter agreeing not to recommence South Africa's space launcher development until South Africa joined the Missile Technology Control Regime (MTCR), which it did in September 1995. Many of the production facilities for fuel and assembly were destroyed or dismantled, but South Africa did hold in reserve an impressive space program infrastructure still in place today. This includes: (1) the Council for Scientific and Industrial Research, with various aerodynamic and materials testing facilities; (2) the Satellite Applications Centre, used for advanced telemetry abilities; (3) the Denel Overberg coastal space launch facility; and (4) numerous industrial aerospace and software producers.

The fall of apartheid and South Africa's transition to democracy in 1994 also helped to establish the legal and institutional framework for a civilian space program, though the difficulties of navigating post-apartheid politics would delay its implementation. The African National Congress came to power with a science and technology policy already formulated, and many of the existing international space treaties, such as the Outer Space Treaty, were finally signed by the new democratic government. Later in the decade, in 1999, South Africa introduced its first indigenous satellite, *SunSat*, designed and developed by graduate engineering students at Cape Town's Stellenbosch University. The 60-kilogram microsatellite was sent into LEO atop an American Delta II launcher and provides low-cost but high-resolution images of South Africa. More than 15 years after it ended, South Africa officially resurrected its space program, but with a radically different mission.

In March 2009, President Kgalema Motlanthe signed into law a bill creating the South African National Space Agency (SANSA). The SANSA budget for 2011 was R93 million (US$12.7 million).[64] In sharp contrast to its national-security-oriented predecessor, SANSA's stated mission is decidedly developmental in orientation and explicitly focuses on the peaceful uses of outer space, such as astronomy, Earth observation, communications, and navigation, and the promotion of international cooperation in space.[65] Moreover, the new space policy is characterized by greater public transparency than its predecessor and, given South Africa's ongoing economic challenges, the fostering of bilateral and multilateral partnerships with Algeria, Kenya, and Nigeria to collectively use remote-sensing data.[66] According to South African Science and Technology Minister Naledi Pandor, "South Africa aims to become a regional center for space technology, investing in satellite and telescope projects to support its ailing economy."[67] Later in 2009, South Africa's second indigenously produced satellite, *SumbandilaSat* ("Lead the Way" in the Venda language), took off from Kazakhstan aboard a Russian *Soyuz-2* launcher. The 81-kilogram microsatellite is designed to track climate change, take reconnaissance photos for agricultural applications, and monitor weather for the southern half of Africa.

South Africa's aspiration to be an African leader in space technology and space research shows much promise. For example, South Africa already accounts for 64 percent of all published science research in Africa.[68] Cape Town's Stellenbosch University founded the country's first satellite engineering program in 2005, and in 2006 the South African government designated US$270 million to build a network of radio dishes to attract the proposed Square Kilometer Array (SKA), a massive collection of radar dishes that would be 100 times more powerful than any current radio telescope. This investment is three times the current annual budget of the South Africa's National Research Foundation, giving some notion of the bet that the South African government is placing on its new space policy. South Africa is already home to the 11-meter Southern African Large Telescope, which began operations in 2005 as the largest optical telescope in the southern hemisphere and on the African continent. Government supporters of such large investments in "pure science," such as the SKA, submit that the new program will contribute to the region's human and technological development by building capacity in engineering and information technology, and will help to inspire young Africans to study science.[69] On the industrial front, a number of South African companies, including Aerosud and Denel Saab Aerostructures, are already aerospace contractors to the Airbus consortium and aspire to be global aerospace suppliers for South Africa's future in space.[70]

Iraq: hopeful Babylon redux

Finally, Iraq provides an interesting instance of a second tier country whose emerging space program was at first hampered by UN sanctions and then ultimately cut short by outside forces—specifically, the American invasion of 2003. Iraq's road to develop a space program was emphatically driven by its perceived security interests. While Iraq was a signatory of the 1968 Nuclear Non-Proliferation Treaty, the country's vice president at the time, Saddam Hussein, secretly ordered the formation of a nuclear weapons program in the early 1970s. These weapons were, in due course, to be deliverable by a ballistic missile.[71] Iraq consistently sought to develop an effective long-range weapons delivery capability, and its acquisition of *Scud* missiles in the 1970s and 1980s was an important first step. But as the Iran–Iraq War ground into trench warfare during the 1980s, the Saddam Hussein government sought a way to break the deadlock and turned to improving Iraq's missile capability.

In 1985, Iraq began to produce solid-propellant rocket motors and to test them through a program called Project 395.[72] At the same time, to improve its indigenous design, Iraq entered into an agreement with Argentina to acquire its *Cóndor II* missile (called the BADR-2000 in Iraq).[73] The Iraqi Ministry of Industry and Military Industrialization assumed control of the program with the public goal of placing a 100-kilogram payload into a medium orbit (200–500 kilometers in altitude). This was a cover for Iraq's real program, the newly dubbed Project S-13, which sought to develop the nuclear-tipped missile.[74] Iraq began construction of its own assembly facilities for the *Cóndor II* in 1987, and although the program ended due to disputes between Iraq and Argentina, by the last year of the war in 1988 Iraq had nonetheless acquired sufficient missile capability to retaliate against Iran. Almost 200 of the improved missiles (now named *Al Hussein*) were fired against Iranian cities. In the same year, Iraq's space program began with the official stated mission of putting satellites into low Earth orbit for reconnaissance and telecommunications.

The space project began inauspiciously with Project Babylon, a plan to build three "superguns," the largest to be over 150 meters long and able to put 200-kilogram payloads into orbit. A smaller 45-meter version of the supergun with a 350-millimeter barrel was completed in Jabal Hamrayn, 145 kilometers north of Baghdad. Permanently mounted on a hillside at a 45-degree angle, the gun was also envisioned as a possible anti-satellite weapon.[75] The project was halted after the Canadian engineer behind the project, Gerald Bull, was assassinated in Brussels. But the 1991 Gulf War affected Iraq's efforts by diverting research toward a modified version of the *Cóndor II* as a satellite launcher from the Al-Anbar Space Research Center, a complex of some 70 buildings about 90 kilometers west of Baghdad.

Within a few years, the Iraqi government was ready to wade into the ocean of space. In 1987, Iraq ordered the production of a more advanced three-stage launcher that would be able to place a small satellite in orbit. This launcher system, secretly called Project S-19 (or alternately *Al-Abid*), was however in reality a test platform for Iraq's long-sought-after nuclear delivery system.[76] In 1989, just one year after Israel had demonstrated its growing space competence by putting its *Ofeq* satellite into orbit, Iraq launched the three-stage *Al-Abid* rocket using five SCUD-Bs clustered together for the first stage.[77] Instantly spotted by the North American Aerospace Defense Command (NORAD), the rocket had no separate satellite for orbit, but instead used its third stage as a test satellite, which completed six orbits before it burned up on reentry. The successful test made Iraq the ninth country to have achieved an independent orbital space launch (see Table 4.2).

But following Iraq's defeat in the 1991 Gulf War, the terms of the cease-fire, as dictated by Resolution 687 of the UN Security Council, established a commission (UNSCOM) to oversee the dismantlement of most of Iraq's missile program. Iraq was limited to purchasing or manufacturing missiles with a range no greater than 150 kilometers. Iraq's remaining missiles were destroyed by inspectors of the United Nations Monitoring, Verification and Inspection Commission (UNMOVIC). In addition, the Al-Anbar center was damaged in the war, effectively halting Iraq's ability to continue its small-scale space race with Israel.

But despite these setbacks, Iraq restarted its missile improvement programs in violation of UN mandates. Iraq attempted to work around the 150-kilometer limitations of UNSCR 687 by developing a shorter-range

Table 4.2 First independent orbital launches

Soviet Union	1957
United States	1958
France*	1965
Japan	1970
China	1970
United Kingdom*	1971
India	1980
Israel	1988
Iraq	1989
Ukraine	1992
Iran	2009

Note
*France and the UK no longer launch independently, but as part of the ESA.

missile, originally dubbed *Ababil*, in which they sought to perfect a liquid-fuel design.[78] After expelling UN inspectors in 1998, Iraq began work on an associated project to create a cruise missile.[79] Tests began in 2001 on an improved launcher, the *Al Samoud II*, which exceeded UN range mandates and left room for speculation as to where Iraq might have progressed in the coming decade. But the toppling of the Iraqi government in the 2003 US invasion ended a promising, if nefarious, space program. The post-invasion Iraqi government sent a letter to the UN Security Council in 2010 declaring its intention to sign the Hague Code of Conduct against Ballistic Missile Proliferation, and the Iraqi government has no current plans to resurrect the launcher program in the foreseeable future.

5 Third tier space actors

I have learned to use the word "impossible" with the greatest caution.
Wernher von Braun[1]

Constructing a definition to classify developing states belonging to the third tier of emerging space actors is more problematic than for first and second tier EMSAs. For the former group of countries, the definition of national security now transcends the traditional realist paradigm to include a plethora of socioeconomic and political benefits derived from ownership of a slice of the space pie. Quite a few developing states today utilize space-based assets for a wide variety of applications, including communications, weather monitoring, and resource planning, even if the data is merely purchased from more developed space actors.

Thus, almost any developing country with a policy toward creating and/ or using space assets, and which does not have the capabilities to be categorized in the first two tiers, would by definition be a constituent member of the third tier. These states invest in space-based technologies while not necessarily possessing even rudimentary launch capabilities, indigenous space industries, or even an official space agency. Those few third tier countries that have achieved launch capability have been restricted— either because of funding limitations or international relations dynamics— to sounding rockets for scientific experiments and, therefore, hampered (for the time being, at least) in their ability to place satellites into orbit.

It may be tempting to assume that the space programs of third tier states exist merely for prestige's sake, given that the level of human development in most third tier countries ranges from medium to low in the 2010 United Nations Human Development Report. Accordingly, the typical criticism lodged against third tier states for their investments in space activities is that such outlays would be better spent on economic and social projects that would yield more immediate and tangible results. But, in the place of traditional military-oriented security concerns, third tier EMSAs have for the most part genuinely embraced space activities as legitimate long-term means to promote socioeconomic development, though garnering some prestige—especially for nationalist purposes—in the bargain has never been eschewed. Much as many lesser developed countries have embraced cell phone technology to leap-frog the cost barrier of landline phones, many third tier EMSAs now look to space-based technology to similarly benefit their societies in an expedient and cost-effective fashion. The countries occupying the third tier of space programs are as geographically diverse as their respective capabilities.

For the purposes of this limited study, only a representative group of third tier states with formal government space agencies will be examined. For some of the more advanced third tier countries, the pursuit of the MNS triad has at some point influenced the evolution of their respective space policies. But, for a variety of reasons, most of these borderline but aspiring space actors have either abandoned, or at least suspended, their more grandiose ambitions. Even in the absence of a more traditional security-oriented space policy, however, these third tier space actors which strive to better their societies through space-based technologies nonetheless fit squarely within the realist realm of competitive self-interest, even as the justification for a state's space policy escapes the orbit of classical hard power. In none of the cases examined herein do states engage in space activities purely for sake of cooperation in the mold of international liberalism. Instead, their space policies are crafted to improve their respective states' socioeconomic well-being as well as to occasionally bring greater independence (and possibly prestige) to the state. In some cases, it merely comes down to the economic calculus that a more independent space policy will yield financial savings that can presumably be employed in other areas of development. In other words, they look to space to improve their societies and provide multiple layers of security—economic, social, and even occasionally military. So while these newer third tier space actors may not have the technology or multi-use aspirations of a China, India, or even Iraq, their space policies are still servicing the perceived national good. We will examine these third tier space actors region by region, beginning with the most advanced state in this tier, Argentina (see Table 5.1).

Argentina's forking paths

In his 1941 short story "Garden of the Forking Paths," the Argentine author Jorge Luis Borges imagined a labyrinth that folds back upon itself in infinite regression, to encourage readers to conceive of all possible choices. Borges's imaginative world fittingly parallels his native Argentina's venture into, out of, and back into the development of a national space program. Among third tier space actors, Argentina is arguably the most advanced, and its situation is illustrative of both the ambitions and the limitations that many similarly aspiring space actors share.

Although at one time Argentina possessed both ambitions and the nascent technology that would have eventually made it a more advanced space actor, the government of post-military-rule Argentina (after 1982) publicly rescinded the country's vaunted goals of developing advanced missile and nuclear weapons technologies. The irony in this rejection is that Argentina had been an early leader in the developing world in the growth of the MNS triad, having built one of the world's first jet fighters (see Chapter 2). Immediately after the end of World War II, Argentina's leaders expressed interest in and even financial support for the development of nuclear weapons and ballistic missile technologies.

Formal interest in rocketry and space flight in Argentina can be traced to the establishment of the Sociedad Argentina Interplanetaria (Argentine Interplanetary Society) in 1948, which was the first private space organization in South America and a founding member in 1951 of the International Astronautical Federation.[2] The group's leader, Teofilo Tabanera, led the campaign for Argentina to establish a state-sponsored space organization, which took form in 1960 with Tabanera becoming the director of the Comisión Nacional de Investigaciones Espaciales (National Commission of Space Research—CNIE). The CNIE worked with the Argentine Air

Table 5.1 Space agencies of Latin America

Country	Space agency	Current space activities
Argentina	National Commission of Space Activities (CONAE)	• satellite production • satellite control • bilateral cooperation
Bolivia	Bolivian Space Agency (ABE)	• satellite control
Brazil	Brazilian Space Agency (AEB)	• launch facility • satellite production • satellite control • bilateral cooperation
Chile	Chilean Space Agency (ACE)	• bilateral cooperation
Colombia	Colombian Space Commission (CCE)	• satellite production • satellite control
Mexico	Mexican Space Agency (AEXA)	• satellite production • satellite control
Peru	National Commission of Aerospace Research and Development (CONIDA)	• launch facility (sounding rockets) • satellite production • satellite control • bilateral cooperation
Uruguay	Aeronautics and Space Research and Dissemination Center (CIDA-E)	• bilateral cooperation
Venezuela	Bolivarian Agency for Space Activities (ABAE)	• satellite control • satellite production (early stages)

Force's Instituto de Investigaciones Aeronáuticas y Espaciales (Institute of Aeronautical and Space Research—IIAE). With the two agencies working together, a number of high-altitude sounding rockets were built and flight-tested to 500 kilometers in altitude, making Argentina the first Latin American country to send a domestically produced rocket into space. The IIAE also produced the first liquid-fuel launcher in South America. At the same time, the Argentine government began its own missile program, building small liquid-fuel rocket motors at the Instituto Aerotécnico in 1947 and solid-fuel engines in 1954.[3]

Concurrent with its early rocket experiments, Argentina established its Comisión Nacional de Energía Atómica (National Atomic Energy Commission—CNEA) in 1950, spurred in part by persistent tension and competition with neighboring Brazil. By 1967 Argentina was operating three nuclear research reactors. Jealously guarding its national sovereignty as well as its strategic options, Argentina openly rejected international calls to join the Latin American nuclear non-proliferation Treaty of Tlatelolco or the broader Nuclear Non-Proliferation Treaty in 1968.[4] Argentina's nuclear efforts were aided generously in matériel and technology from Canada, West Germany, Switzerland, and the Soviet Union. By 1983, all indications were that Argentina was inching toward the development of an indigenous nuclear weapon, evidenced in part by the country's well-developed nuclear infrastructure at the Balseiro Institute, whose research reactor in southern Argentina was capable of enriching weapons-grade uranium. Consequently, Argentina became an important secondary seller of nuclear technology and related high-tech equipment to other developing and aspiring regional powers, especially in North Africa and the Middle East.[5]

The growth of space and missile programs in Argentina accelerated between 1961 and 1975. On its own, the Argentine air force's Grupo de Desarrollos Espaciales (Space Development Group) initiated a domestic sounding program during the early to mid-1960s for measuring x-rays and upper atmospheric temperatures using the indigenous, two-stage *Gamma Centauro* rockets. Launched from sites on the Argentine mainland as well as the Matienzo military base in the Argentina-claimed portion of the Antarctic Peninsula, these rockets gathered basic scientific data and also functioned as a technology template for future missile development.[6] Concurrently, Argentina also participated in a number of cooperative agreements with NASA on sounding-rocket programs. In all, Argentina's National Space Research Commission (CNIE) and the Institute of Aeronautics and Space Investigations (IIAE) developed and tested 11 different rocket designs during the 1960s.[7] The last sounding rocket, the two-stage *Rigel*, was launched seven times from 1967 to 1973 and ascended to an altitude of over 300 kilometers.

After much experimentation, the final missile design to emerge during the 1970s from the Argentine air force's General Management of Space

Projects division was for the *Cóndor* missile. Lacking sufficient funds to continue development on its own, Argentina sought the help of the West German company Messerschmitt-Bölkow-Blohm (MBB) to produce a single-stage, short-range tactical missile, though it was publicly labeled a furtherance of the country's sounding rocket program.[8] Other European firms from Switzerland, Italy, and Austria were also contracted to contribute to various aspects of the program. Not coincidentally, the rise of the *Cóndor* project coincided with rising militarism in Argentina just preceding the country's notorious period of political repression, the "Dirty War." During this time of upheaval, through an initiative called the National Reorganization Process, military officers increasingly displaced civilian aerospace engineers in the CNIE in order to promote military projects over civilian space developmental efforts.[9]

Having benefited from US-supplied technical data on sounding rockets and space launch vehicles (i.e., short-range and intermediate-range missiles, respectively), such as the US-supplied *Castor* rocket, Argentina finally constructed its own major missile assembly facilities for the *Cóndor*, located at the then-secret US$200-million Falda del Carmen ("Carmen's Skirt") complex near Córdoba in north-central Argentina. Static tests on the *Cóndor* were conducted, but the project did not ultimately produce a workable ballistic missile. It did, however, lay the groundwork for the later, though never deployed, short-range ballistic missile *Alacrán* ("Scorpion"), which itself served as a test bed for the updated version of the *Cóndor* missile.

After Argentina's ignominious defeat in the 1982 Falklands War (which was assured, in part, by a general arms embargo against the country), as well as ongoing tensions with Chile over three disputed islands in the Beagle Channel, the Argentine military became resolved to create an independent ballistic missile deterrent. Accordingly, Argentina signed a covert agreement with Iraq to develop the more sophisticated two-stage *Cóndor II*, which was to have a 500-kilogram payload capacity and a 1,000-kilometer range.[10]

Argentina's relationship with Iraq was not a unique one. During the late 1970s, Argentina had diversified its contacts, especially in the Muslim world, to facilitate sales to provide continued funding for the project.[11] Clandestinely funneling upwards of US$1 billion through Egypt (which received assurances of *Cóndor* missile technology for its cooperation) from 1984 to 1989, Argentina struck the secret deal with Iraq, which had aspired to gain advanced, longer-range missile technology to attack Israel and Iran, while, in exchange, Argentina's air force hoped to finally acquire a launcher for an anticipated nuclear warhead, all done under the guise of space research. The Argentine government additionally had high hopes of developing, with the help of surreptitious Saudi financing, a lucrative missile market in the Middle East to complement Argentina's already firmly established nuclear energy market (Argentina was the site of Latin

America's first nuclear power plant, Embalse, opened in 1983).[12] The Italian firm SNIA-BPD was secretly contracted to help iron out technical difficulties with the rocket's guidance system. Detailed information on the flight testing of the *Cóndor II* is sketchy at best and contradictory at worst. Some reports claim it was test-launched in 1986 in Patagonia, while others deny that flight tests ever took place.

In the end, despite the vociferous opposition of some Argentine air force officers, the combination of the project's ongoing technical problems, the withdrawal of external financial support and missile orders, US threats to cancel financial credits to the country, and Argentina's own financial meltdown produced a critical juncture.[13] Under the totality of this crushing pressure, the recently elected president, Carlos Menem, discontinued the *Cóndor II* program in 1990, though the official public rationale was that the program was discontinued in accordance with the MTCR. In a dramatic about-face from its previous stance, Argentina also ratified the Treaty of Tlatelolco in March 1993 and became a signatory of the Nuclear Non-Proliferation Treaty in 1995. But while publicly axing Argentina's military missile programs, Menem re-stoked the fires of the new space program in 1994 by approving a generous US$700-million space budget.[14] The difference was that now the unnamed space program was to be civilian and serve Argentina's scientific and socioeconomic ends.

Under Menem, other institutional changes to the space program occurred. Replacing the previous military-run missile and space programs, the National Commission of Space Activities (CONAE) was founded in 1991 to capitalize on the *Cóndor II* program's advances within a civilian context and to refocus the program's scientific advances toward remote-sensing and scientific satellites. Since this reorganization, in cooperation with Brazil, France, Netherlands, and Italy, Argentina has so far produced a variety of space applications and become one of the developing world's most active collaborators in satellite projects.

Argentina's first home-built satellite, *MuSat-1*, was a US$1.2-million government-sponsored project of the Instituto Universitario Aeronaútico de Córdoba that tested low-cost communications and imaging technologies. It was launched in 1996 from Russia's Plesetsk Cosmodrome aboard a Molniya orbital launch vehicle. Argentina's next effort, the mini-satellite *SAC-A* ("Scientific Applications Satellite-A"), was placed into orbit by the space shuttle *Endeavour* in December 1998. Intended for rudimentary GPS services and remote imaging, it was built by Argentina's state-run, high-tech company Invap S.E., a NASA-qualified (i.e., pre-approved to work with NASA) company that also produces small nuclear reactors and has provided related nuclear technologies for Argentina as well as over 30 other countries.[15]

In the past decade, Argentina has taken an important step toward utilizing its space achievements to further the country's infrastructure and socioeconomic development. In 2004, the Argentine Senate approved the establishment of a national satellite company called ARSAT (Argentina Satellite Solutions Company). This company was made responsible for producing a satellite to fill the valuable 81-degree orbital slot that had been assigned to Argentina in 1998 by the International Telecommunications Union. With a US$16-million budget, ARSAT was to absorb the rival multi-national consortium Nahuelsat (composed of Aerospatiale, Daimler-Chrysler, and Alenia Spazio), leaving Argentina with a single domestic satellite producer. A series of three medium-sized geostationary communications satellites is planned under the ARSAT name, with launches scheduled for 2012, 2013, and 2014. The ARSAT satellites will provide data, telephone, and television services to most of South America (except Brazil). However, their launch will still be dependent on outside services; the first in 2012 is scheduled to be launched aboard an *Ariane 5* from French Guiana.[16]

But to fill the missing link in the space chain, a domestic launcher, Argentina began discussions in 1997 with its erstwhile rival Brazil to jointly design and build a research sounding rocket called VS-30. In 1998 the two hopeful South American space actors signed a bilateral accord for space cooperation, which in 2007 resulted in a successful suborbital rocket launch from Brazil's Barriera do Infierno test site on which both countries conducted microgravity experiments. This was the first successful collaboration between two Latin American countries on a space launch. Also in 1998, Argentina initiated an international satellite data collection point through its Centro Espacial Teófilo Tabanera ground control station in Córdoba, which receives data from a dozen different international satellites. A second receiving station, dedicated to Antarctic data collection, is planned to be built by 2012 in Ushuaia, in the Tierra del Fuego province at the southern tip of South America.[17]

However, like so many smaller space actors, Argentina's growing space ambitions have necessarily attracted an influx of foreign investment. In Argentina's case, recent years have brought a deluge of Chinese investment in both its telecommunications and space industries. Argentina's 2001 economic implosion left an investment vacuum into which Chinese companies rushed. While Argentina and most other Latin American countries have historically relied disproportionately on US support and technology for their space activities, that situation began to change following the Cold War. While the US continued to launch Argentine-built satellites through the 1990s, China made a formal approach to cooperation through a 2004 agreement, "Technology Cooperation in the Peaceful Use of Outer Space," which placed China in a privileged position to provide Argentina with commercial launch services, satellite components, and communication satellite platforms, thus diluting Argentina's historical space technology relationship with the United States.[18] But this has not put an end to the US relationship, only diversified it. CONAE has continued cooperative satellite projects with NASA, such as the recently launched collaboration called SAC-D/Aquarius, which was launched by a *Delta II* launcher from California's Vandenberg Air Force Base on 10 June 2011. Housed aboard an Argentine SAC-D satellite, its ground-breaking mission is to measure ocean salinity in order to better understand climate change.

In addition to fomenting cooperative programs with Brazil, Argentina has actively sought other avenues of international space cooperation, such as through a cooperative agreement with the Italian Space Agency (ASI) to construct the Italian-Argentine Satellite System for Emergency Management, which will be realized via Argentina's proposed SAOCOM (Argentine Microwave Observation Satellite) satellites, expected to be launched by 2013 from the United States' Vandenberg Air Force Base. The Belgian space agency will also contribute radar satellite technology for this project.[19] Planning for a launcher phase in the space program began in 1998 with President Menem's establishment of the company VENG (Spanish acronym for "New Generation Space Vehicle"). In 2000, CONAE, which runs VENG, started to plan the financing and engineering of Argentina's new space launcher, which is designed to be capable of putting light payloads of communications and remote-sensing satellites into low Earth orbit.[20]

In 2005, the late Argentine President Néstor Kirchner signed a decree which declared the promotion of space technology to be a state policy and a national priority. Argentina's national space program was formally institutionalized by Kirchner via Presidential Decree 350 in April 2007, which calls on all state-sponsored scientific and engineering organizations, including CONAE, state universities, the Argentine Aeronautical Institute, and the Institute of Technical and Scientific Research of the Armed Forces (CITEFA), to work jointly toward the stated goal of building domestic launchers and providing access to space, particularly light cargo satellite services. The pronouncement was, in effect, an open call to wake up the Argentine space program from its decade-long slumber. In the same year the Italian *Cosmo-Skymed II* Earth-observation satellite was launched from Vanderberg Air Force Base in California, and Argentina was one of the major participants in the control of the satellite through its control station located in Córdoba.

The primary reason for the development of Argentina's domestic launcher, according to CONAE director Conrad Varotto, is that Argentina cannot continue to depend on the US or other countries to get Argentina's satellites into space and that the high launch costs of "acceptable" providers make the development of an Argentine launcher the least costly alternative.[21] The most important step in Argentina's policy of space self-sufficiency is underway in the development of Argentina's first light-payload satellite launch vehicle for domestic and commercial purposes. The new launcher has been dubbed *Tronador* ("Thunderer," after an extinct glacier-covered stratovolcano on the Argentina–Chile border), and is reportedly a completely civilian project under CONAE. However, the participation in the project of the Bariloche-based Balseiro Institute, a renowned nuclear engineering school responsible for the basic research in the country's defunct *Cóndor II* missile and the former nuclear weapons programs, makes the presumption of no possible future military applications for *Tronador* somewhat problematic.[22]

A number of universities, such as the Instituto Balseiro Universidad Nacional de Cuyo, have contributed to the design.[23] The first fruit of the renewed efforts was the four meter-high, liquid-fuel *Tronador I* prototype that was test-flown in July 2007 at Bahía Blanca, home of South America's largest naval base, located 180 kilometers south of Buenos Aires. The working version, the two-stage, 33-meter-tall, 64-metric-ton *Tronador II*, is being constructed at the Falda de Carmen facility and is reportedly being designed right up to the limits imposed by the MTCR. Once *Tronador II* becomes operational, its hydrazine-fueled engine will thrust Argentina into second tier status.[24] Capable of lifting up to 250 kilograms into LEO at an altitude of 400 kilometers, *Tronador II* has the potential to position Argentina as a viable competitor in the light commercial satellite launches.[25] Its test flight is scheduled for the end of 2012.[26]

More broadly, the *Tronador* project addresses one of the principal goals of Argentina's current "National Space Plan: Argentina in Space, 2004–2015," which declares that the space program will be essential in the improvement of six areas: (1) agriculture, fisheries, and forestry; (2) climate, hydrology, and oceanography; (3) disaster monitoring; (4) environment and natural resource monitoring; (5) cartography, geology, and mining; and (6) health.[27] It further states that the space program is charged with a mission to "sense, collect, transmit, store, and process information relating to economic and production activities, the environment, and geophysical characteristics of the continents and oceans of our planet and, particularly, [Argentina's] national territory."[28] As a matter of national development policy, space activities now have a privileged place at the table in Argentina's national agenda.

Other Latin American space actors

The rest of the Latin American countries with an interest in space-based activities have committed themselves to some level of research, satellite investment, and/or cooperative endeavors in space-related activities. Central American and Caribbean countries, with their relative poverty and limited resources, have not expressed an interest in pursuing space-related projects, while Mexico and many South American countries have begun the process of building small, targeted space programs. While the typical project involves the purchase of telecommunications satellites, a few have shown the incipient signs of aspiring to emulate the more advanced space actors through embryonic sounding rocket programs and related aerospace research.

Regardless of these ambitions, with the exception of Brazil, the level of space technology in Latin America is still comparatively in its infancy and is like to remain so for the foreseeable future because of the costs, relative lack of space technology infrastructure, and persistent socioeconomic challenges. Despite these obstacles, the seeming chasm of challenges to access space has not deterred some of the more determined Latin American states from pursuing some measure of space-related activities. In every case, the motivation for their investment in space remains constant: the cost-benefit relationship of space assets has been deemed advantageous both for reasons of socioeconomic gains and, to varying extents, the prestige earned by joining the space club. Also, for a few states that exhibit the unrelenting presence of military prerogatives, the emerging space policies of their countries provide yet another opportunity to address long-standing institutional needs.

Mexico: so close to the US, not so far from space

Mexico's late nineteenth-century dictator Porfirio Díaz quipped that Mexico was "so far from God and so close to the United States." Despite the implied geographical disadvantage and the very real tensions that have ensued between the two countries since the Mexican–American War (1846–48), Mexico's proximity to the United States has nonetheless been advantageous to Mexico's space policy ambitions, allowing it to benefit from its more space-savvy northern neighbor. Mexico is a typical third tier EMSA in that it currently lacks independent launch capacity but nonetheless pursues an increasingly active space policy.

Whereas rocket experiments by private Mexican citizens had occurred earlier in the twentieth century, the first semi-official rocket experiment took place in December 1957 at the Escuela de Física Gustavo del Castillo in the state of San Luís Potosí. Under the tutelage of the school's director, Dr. Gustavo del Castillo y Gama, a group of students constructed a 1.7-meter sounding rocket that reached an altitude of 2,500 meters for the purpose of conducting measurements of cosmic rays.[29] Mexican government-sponsored experiments with advanced rocketry followed. With rocket engines based on the German V-2, Mexico's two separate four-meter-tall, 200-kilogram sounding rockets, the SCT-1 and the SCT-2, were launched into LEO for weather experiments in 1959 and 1960 respectively.[30] In 1962 the Comisión Nacional del Espacio Exterior (National Commission of Outer Space—CONEE) was established by the decree of President Adolfo López Mateos as a section in the office of the Secretariat of Communications and Transportation. Concomitantly, the National Autonomous University of Mexico established an academic division called the Department of Outer Space (today renamed "Department of Space Sciences").

Mexico's modest yet globally recognized next step into other space activities began as a sport—literally. In order for the 1968 Mexico City

Olympic Games to be the first televised in color, the CONEE constructed a ground station to transfer the satellite signal, which reflected a larger trend in the region. By 1968 almost 300 Latin Americans in Mexico, Chile, Ecuador, and Peru were working at tracking stations that had been set up through cooperative projects with NASA as part of the US space effort.[31] In the same year, the government-run Satmex company was formed to process satellite signals and Mexico became a signatory of the International Telecommunications Satellite Organization (INTELSAT).

Besides the obvious technological gains, Mexico's space policy fulfilled three political goals as well. First, it helped bolster the country's image as a potential technological leader in Latin America. Second, Mexico's sounding rocket program served as an example of the country's increasing level of modernity (similar to India's motivations). Third, it served to reinforce diplomatic relations with the United States, which led to future cooperation on several early NASA space projects, including the first US manned program, Project Mercury, for which a tracking station was built in Mexico.

The evolution of Mexico's space activities was not smooth, however. The CONEE was dismantled by President José López Portillo in 1977 because of a lack of funding and political support. The unfortunate result was that a significant portion of Mexico's growing community of aerospace engineers and those holding doctorates in aerospace-related sciences left the country to find work.[32] Despite this stumble, Mexico was one of the first developing countries to adopt a space policy that put satellite technology at the forefront of socioeconomic development.

Mexico's first two telecommunications satellites were built by the Hughes Corporation. The *Morelos 1* (1985) and *Morelos 2* (1998) were placed into geostationary orbit from the space shuttles *Discovery* and *Atlantis*, respectively. The latter mission also included Mexico's first astronaut, payload specialist Rodolfo Neri Vela, who conducted human physiology experiments. Five more satellites have followed, including the amateur radio satellites UNAMSAT-A and B, built by students from the Autonomous University of Mexico. UNAMSAT-A was lost in 1995 after launch when the Russian *Start-1* launch vehicle failed to achieve orbit, but UNAMSAT-B was successfully placed in orbit from Russia's Plesetsk Cosmodrome on 5 September 1996. Mexico plans to expand its space assets with the launch of three new Mexsat satellites in 2012–14. Costing about US$1.5 billion and being built by Boeing Space Systems, these satellites will include a telecommunication platform for the health and education sectors as well as Mexico's first dedicated military reconnaissance platform.[33] While it has yet to build its own satellites, Mexico has invested heavily in satellite command and control systems as well as building the basis for a strong aerospace sector. As of 2009, over 232 aerospace companies, such as Honeywell Aerospace and General Electric, operate in Mexico and generate almost US$500 billion annually in exports.[34]

After debating the matter for years, on 31 July 2010, the Mexican government approved the creation of the country's official space agency, the Agencia Espacial Mexicana (Mexican Space Agency—AEXA). The AEXA is charged with developing Mexican space policy, stimulating investment in aerospace technology, and coordinating all Mexico's space activities, from basic scientific research involving the country's ground-breaking Large Millimeter Telescope to aerospace construction activities. In addition, AEXA will oversee a proposed US$80-million launch facility to be constructed near the city of Chetumal in an unpopulated region of the Yucatan Peninsula.[35] However, with an initial operating budget of only US$800,000, AEXA will be hard pressed to accomplish much of its mission, and autonomous launches from Chetumal are not expected for a decade or more.[36]

Peru: the Inti revisited

Peru holds a special place among Latin America's EMSAs because the country was home to Pedro Paulet, who invented the world's first liquid-fuel rocket engine in 1895 and the first modern rocket propulsion system in 1900. Though Paulet's invention was never attached to a working rocket, his invention is recognized as being decisive in the furtherance of space flight. According to Wernher von Braun, "Paulet should be considered the pioneer of the liquid fuel propulsion motor ... by his efforts, Paulet helped man reach the moon."[37] Paulet went on to found Peru's National Pro-Aviation League, a precursor of the Peruvian Air Force.[38]

The country's more modern dabbling in space activities came after a period seeking to acquire ballistic missiles for more down-to-earth strategic reasons, namely Peru's long-standing territorial dispute with Ecuador, which culminated in three military conflicts in the twentieth century. During the 1970s, Peru was one of at least 20 developing countries seeking to acquire ballistic missiles.[39] Peru's first experimentation in space projects came with its participation in NASA upper atmospheric experiments in 1975 (Project Antarqui), and equatorial magnetic field measurements in 1983 (Project Cóndor), both conducted at the Peruvian air force base Punta Lobos, located in Pucusana, 50 kilometers south of Lima.

Peru's Comisión Nacional de Investigación y Desarrollo Aeroespacial (National Commission for Aerospace Research and Development—CONIDA) was founded in 1974 to be a civilian research organization for space activities, conclude collaborative agreements with foreign space organizations, and support Peruvian national space projects.[40] However, the Peruvian military has always had a very close relationship with CONIDA, from chairing the organization to providing the research and launch facilities. In recent years, CONIDA has worked to establish more and deeper ties with more established space actors as well as other EMSAs to advance Peru's own abilities and options.

In 1998 Peru signed a memorandum of understanding on space coop-eration and technical assistance with the Indian Space Research Organisa-tion and a similar agreement with the Russian Federal Space Agency (RKA).[41] However, despite these and a number of other international agreements aimed at cooperative space activities, none has yet produced a tangible project. Concurrently, the Fujimori administration entered into negotiations with North Korea to purchase that country's *Scud-C* missiles, presumably to deter Ecuador from further incursions along the two coun-tries' disputed border but also as a form of technology transfer. The deal never materialized because of the Peruvian government's concern for the destabilizing effect the missiles might have in the region and for fear of US economic sanctions.[42]

A project called CONIDASAT was begun in the early twenty-first century to produce an indigenous remote-sensing satellite that would address the problems of adequately covering Peru's challenging geography, which is divided between coastal deserts, the Andes mountains, and the Amazon jungle. The project advanced far enough that a clean room for satellite assembly was built; however, the project was finally cancelled in 2003 for lack of funds. Nonetheless, the experience began the process of training Peruvian engineers in the intricacies of satellite design and construction. In October 2005, Peru took concrete steps toward advancing its space program by becoming the only Latin American member of the Asia-Pacific Space Cooperation Organization (APSCO).

The following year saw multiple advances in Peruvian space policy. In July 2006, the Peruvian congress approved Law 28799, which declared the coun-try's "national interest in the creation, implementation, and development of a 'National Satellite Imaging Operations Center'."[43] In December of the same year, Peru launched a 2.7-meter, 99-kilogram sounding rocket, *Paulet I*, from the Punta Lobos. The rocket carried aloft devices to measure condi-tions in the upper atmosphere, including pressure, temperature and humid-ity, and astrophysics equipment. CONIDA's director, Peruvian Air Force Colonel Wolfgang Dupeyrat Luque, declared that with the launch "[Peru] enters a new era of aerospace development" and that Peru seeks to develop its own space program.[44] In spite of the civilian, developmental face of CONIDA, the leadership of Peru's space agency by a military officer reflects the long-standing institutional interests of the Peruvian military in a future space program in the country. Launched in September 2009 was the follow-up sounding rocket *Paulet II*, which reached an altitude of 20 kilometers.

To support these expanding satellite projects and in fulfillment of Law 28799, the Peruvian government established the National Committee for Operations of Satellite Images (CNOIS), whose mission is to promote technological and scientific development in Peruvian remote-sensing operations, especially using advanced but inexpensive nanosatellites and picosatellites, which weigh less than 10 kilograms and one kilogram, respectively.[45] The first domestic Peruvian satellite attempt is the

10×10×10-centimeter, remote-sensing nanosatellite called *Chasqui-I* ("Messenger-1" in Quechua), which is scheduled to be launched by Russia in 2012. Based on the Cubesat design originally created at Stanford University (USA), *Chasqui-I* was built by students at Peru's National Engineering University to provide practical experience in satellite technology to the country's future engineers. The nanosatellite will take photos of Peru from an altitude of 600 kilometers. The successor, *Chasqui-II*, is currently being built in cooperation with the ESA, and a third satellite is slated for completion in collaboration with Russia's University of Kursk.

In 2009, Peru issued its first national space policy, La Política Espacial del Perú (The Space Policy of Peru), in which space activities are officially identified as exclusively an instrument of sustainable development, thus publicly and formally eschewing the interests of the Peruvian military. Both public and private entities are called on to contribute to Peru's space future because "without science and technology the country cannot achieve development ... and without development there is no security."[46] Interestingly, the policy document directly recognizes the relationship between development and space activities by noting that the major developed countries also possess space programs as well as pointing to the beneficial effect space programs can have in terms of sovereignty, political and economic power, and military capabilities, in addition to numerous socio-economic advantages including the investigation of global climate change.[47]

Ecuador's modest hope

Peru's northern neighbor and erstwhile foe, Ecuador, has also embraced space activities as a tool for national development. Ecuador's experience with space programs has been ongoing for over 50 years. In 1957, Ecuador was one of the sites for NASA's early satellite control stations, situated at the foot of the 5,897 meter stratovolcano Cotopaxi. In 1977, the Center for National Resource Extraction by Remote Sensing (CLIRSEN) was formed to aid in the country's burgeoning petroleum industry. In 1982, CLIRSEN took over operational control of NASA's Cotopaxi stations when their mission ended, and transformed them into stations for national satellite data processing.

Ecuador's space prestige moment arrived in 2003 when Ronnie Nader Bello, an Ecuadorian systems engineer, became the first Ecuadorian to receive training as an astronaut at Moscow's Gagarin Cosmonaut Training Center in 2007. Though he has yet to go into space, upon Nader Bello's return to Ecuador, his national fame as a trained astronaut helped him to found the Agencia Espacial Civil Ecuatoriana (Ecuadorian Civilian Space Agency—EXA), a unique mixed capital non-profit organization. However, contrary to its name, EXA is only partly civilian, as the Ecuadorian military (Fuerzas Armadas del Ecuador—FAE) owns a 50 percent stake in the organization, following the FAE's long-tradition of owning businesses in the private sector (such as the national airline, TAME) to supplement its budget.

The stated objectives of EXA are to take place in three phases. The first entails orbital flights of at least 100 kilometers to conduct at least two experiments. The second flight will involve sending an astronaut to the International Space Station for a minimum of 10 days, and the third—and most ambitious—foresees a lunar mission.[48] Despite the questionable nature of the program's ultimate goals, the stated space policy goal to "inspire an entire generation of people, now children" coincides well with the stated developmental mission of EXA's program to produce a more scientifically aware and literate society.[49] A specific step was taken in May 2008 toward realizing the first phase of the plan, when EXA and the Ecuadorian Air Force unveiled a modified Ecuadorian Air Force Sabreliner T-39, which can mimic microgravity (via an elliptic flight path relative to the center of the Earth) and be a platform for conducting studies requiring low or no gravity. Called Project Daedalus, this aircraft is the first microgravity plane of its kind in Latin America, and places Ecuador's young space program in a position to contribute positively to the country and region's space ambitions.[50]

Venezuela's Bolivarian dream

Since his election as president of Venezuela in 1999, Hugo Chávez has implemented a host of policy innovations meant to modify Venezuelan society, sometimes radically. Through his "Bolivarian Revolution," nationalizations of major industries as well as a plethora of populist social programs have been implemented via a carefully orchestrated plan to instill a sense of autonomy and nationalist pride ("Bolivarian" is in reference to the principal hero of Latin American independence from Spain, Simón Bolívar, who was from Venezuela). The growth of Venezuela's current space policy is simply another facet of Chávez's nationalist policies, although Venezuela's initial interest in space activities predates the current Chávez administration and began within the context of the developmental challenges of the greater Andean region.

During the early 1980s, owning to the extreme developmental challenges presented by the region's geography (e.g., the Andes Mountains, the world's longest and second-highest mountain range), the five countries of the Andean Pact—Bolivia, Colombia, Ecuador, Peru, and Venezuela—conducted satellite technical and feasibility studies and identified a slot in the geostationary orbit for an Andean satellite system. The system was to be called Project Cóndor, but it never saw the light of day because of the dire financial straits suffered in the region during the "lost decade" financial meltdown of the 1980s. But by the mid-1990s, the political and economic landscape of Latin America had improved, and privatization, once anathema to many Latin American leaders, was embraced as the preferred economic reform model. In 1994, the Andean Committee of Telecommunications Authorities (CAATEL), together with the Association of Telecommunications Companies of the Andean Region (ASETA), made the decision to advance the project under the new name Project Simón Bolívar.[51] In 1996, the government of Rafael Caldera requested an orbital slot for the *Andesat* telecommunications satellite, which was wholly owned by private investors from the five Andean Pact countries.

But under Chávez, Venezuelan space activities assumed a decidedly more nationalist character. In November 2005, Chávez decreed the creation of the Venezuelan Space Center (Centro Espacial Venezolano— CEV), a subsection of the Ministerio del Poder Popular para Ciencia, Tecnología e Industrias Intermedias (Ministry of Popular Power for Science and Technology, and Intermediate Industries—MPPCT). The mission statement of the CEV is straight-forward and conforms to the country's populist and semi-authoritarian policies: "to implement the policies of the National Executive of Venezuela regarding the peaceful use of outer space." The institution's name was subsequently changed in 2007 to the ever more nationalist Agencia Bolivariana para Actividades Espaciales (Bolivarian Agency for Space Activities—ABAE) and given the expanded mission of "consolidating program plans and research and development projects that allow the insertion of space technology in the decision-making of the Venezuelan public sector, within a strategy of integration and articulation of national networks framed by international cooperation."[52] The ABAE's specifically defined areas of socioeconomic development include the promotion and development of telemedicine, long-distance education, remote sensing, and telecommunications.

The road toward Venezuela's first satellite is illustrative of the new cooperative paradigm that smaller states just beginning to implement their space policies are embracing. In November 2002, Venezuela resumed talks with Brazil to complete the long-delayed *Andesat*, but the country eventually chose instead to engage a non–Latin American space actor. Venezuela signed a technology transfer agreement with China in 2004, which coincided with 19 major cooperation agreements signed by Venezuela with China, including agreements on preferred Chinese access to Venezuelan petroleum and gas, the expansion of Venezuela's railroad system, a line of credit for Venezuela to purchase Chinese agricultural equipment, and a China-provided joint development fund of US$4 billion.[53] In November 2005, Venezuela signed a contract with China's Great Wall Industry Corporation for the country's first satellite, *Venesat-1* (which was China's first for a Latin American customer).

A telecommunications satellite, *Venesat-1* originally began as an Uruguayan project, *Urusat-3*, which had been repeatedly delayed because of Uruguay's financial difficulties.[54] In an agreement with Uruguay, Venezuela assumed the project, changed the satellite's name, and took over

Uruguay's previously assigned geosynchronous orbital slot at 78 degrees west, located approximately 36,000 kilometers above Ecuador. In exchange, Uruguay received 10 percent of the satellite's bandwidth. Once the project became Venezuelan, China proceeded with a training program for 150 Venezuelan scientists to operate the satellite from the satellite control stations located in the southeastern states of Bolívar and Guárico, and even provided scholarships to 30 Venezuelan graduate students to research their doctoral dissertations in China on aerospace-related topics.

Besides transmitting Venezuelan state television programming, the political intent of *Venesat-1* was made patently clear by President Chávez when he described the satellite's mission as helping to end "private media terrorism," meaning the end of private news outlets' ability to criticize the government.[55] The second satellite in the series, *Venesat-2*, is planned for launch in the latter half of 2012, and long-term plans have been set for replacing both satellites in 2022 and 2024. With a total cost for the project of US$406 million (US$241 million for the satellite itself), Venezuela has become Latin America's second largest space actor (after Brazil) in terms of investment in satellites.[56] Through *Venesat-1*, run by Venezuela's national telecommunications company, the Chávez administration is capable of providing voice, television, data, and internet services to almost all of Central and South America as well as the Caribbean, thus providing a platform for the Chávez's national talk show *Aló Presidente* (previously carried by a Dutch satellite) as well as Chávez's stated plan for greater regional integration and autonomy. The last brick laid in Venezuela's road to greater space autonomy occurred in 2010 when the MPPCT announced that China would assist Venezuela to build the country's first small satellite assembly facility, which will be located in the state of Carabobo and be operational by 2012.[57]

Colombia: steady and organized

Colombia's space program is a typical late-blooming third tier EMSA, with no launch facilities or indigenous space industry to speak of. Unlike some of its South American neighbors, Colombia has never pursued missile or nuclear technology, in part because of the country's long-standing demo-cratic traditions and stability (though there are notable exceptions). Without the typical strategic imperatives, the formalization of a space policy in Colombia is a recent development. In July 2006, President Álvaro Uribe's decree No. 2442 created the Comisión Colombiana del Espacio (Colombian Space Commission—CCE) as a government entity to promote the country's development of satellites and space-based technology for navigation, oceanography, and remote sensing of natural resources. The CCE is administered under the Agustín Codazzi Geographic Institute, and Colombia's vice-president is the titular head of the commission. The founding of the CCE was an integral pillar of Colombia's National

Development Plan 2006–10, which emphasizes the urgency of economic and social development, particularly as a vehicle to rectify the adverse socioeconomic conditions that gave rise to the insurgency group FARC (Revolutionary Armed Forces of Colombia). Among the CEE's objectives are the training of talent in satellite engineering and Earth observation as well as to be a vehicle for the transference of space-related knowledge in the country.[58]

The follow-up white paper to this development plan was 2019 Visión Colombia II Centenario, which outlines the steps to accomplish the country's longer-term development goals, specifically advocating the development of the knowledge and use of space technologies in Colombia via Earth observation.[59] The document called for a Colombian satellite to fulfill the project, which would be carried out in three phases, from the development of home-grown feasibility studies and research on a satellite to the construction and application of geospatial data. The result was *Libertad-1*, a 10×10, one-kilogram nanosatellite constructed at the Universidad Sergio Arboleda in Bogotá for the bargain-basement price of US$250,000. Launched into LEO from the Baikonur Cosmodrome on 17 April 2007, *Libertad-1* carried a digital camera and audio transmitter, and served as an experimental platform to improve Colombia's infant space technology sector during the satellite's short 34-day mission.

Bolivia: poverty reduction by satellite

Despite being one of the poorest and least developed countries in Latin America (60 percent of Bolivians live in poverty), Bolivia is nevertheless following in its neighbors' footsteps in pursuing space technology as a means to socioeconomic development. In February 2010, President Evo Morales signed a decree that established the Agencia Boliviana Espacial (Bolivian Space Agency—ABE), which will be based in the capital, La Paz.[60] The ABE will have only a lackluster US$1-million annual budget, which will be cobbled together from a combination of government funding, foreign loans, and donations. The stated mission of the ABE is to promote technology transfer, human-resource development and the application of satellite-communications programs to education, defense, medicine and meteorology. In addition, the ABE will carry out satellite-based prospecting to search for natural resources, especially natural gas, of which Bolivia has the second largest reserves in Latin America.

The ABE's first assignment is to oversee a planned telecommunications satellite project that is slated for completion by 2013. The US$300-million telecommunications satellite *Túpac Katari* (named after the leader of an eighteenth-century Indian rebellion against the Spanish) will be built by China's Great Wall Industry Corporation, using the basic frame of the Chinese DFH-4 satellite, and will be 85 percent financed by a Chinese bank.[61] China will provide training to ABE personnel, who will work at the satellite's two control stations located in Pampahasi and La Guardia in Santa Cruz department in eastern Bolivia.

This marks a dramatic shift for Bolivia, which had for decades relied on US financial assistance. The project follows on the heels of a similar Chinese-backed project in 2004 by Morales's ideological counterpart, Hugo Chávez. The most immediate benefit of the *Túpac Katari* project will come from simple cost-cutting and traditional nationalism. Bolivian companies currently pay approximately US$10 million annually to foreign satellite operators, and the new Bolivian-owned satellite will cut their costs in half.[62]

Chile: overcoming the military legacy

Chile's approach to space-related activities can be divided into two epochs—the military and post-military years. However, in both, the influence of military prerogatives has remained constant. From the country's colonial days until its democratic opening in 1989, military prerogatives in Chile—as in most of Latin America—meant that the Chilean military enjoyed both a special legal status (*fuero militar*) and a privileged place in the country's society. This unique standing became especially notable during the Cold War, with the Chilean Armed Forces's (FAC) neofascist campaign against communist subversion and its wariness of Argentine aggression into the Beagle Channel (which included a planned but unexecuted Argentine invasion of Chile in December 1978 called "Operation Sovereignty"). While no ballistic missile program drove Chile's space-related goals (though the country is not an adherent to the MTCR), the institutional interests of the FAC did have—and still has—an important effect.

At 3 percent, Chile has Latin America's third-highest military expenditures as a percentage of GDP (after Cuba and Colombia), which is made possible because of the vast revenue provided by copper mining (Chile produces one-third of the world's copper). A 1958 law, and amendments made during the military dictatorship, stipulate that 10 percent of revenues of the state-owned copper company Codelco be transferred into an exclusive reserve fund to be used solely for military purchases, thus insulating and depoliticizing the FAC's budget. By law, the allocation cannot be reduced or diverted, and must be supplemented if revenues fall below US$225 million. During the Cold War the FAC spent the annual windfall of approximately US$180 million on some of the region's most high-tech armaments, such as the US F-16 fighter jet and the German Leopard main battle tank.[63] This privileged budgetary position as well as Chile's long history of bilateral military cooperation with United States and Europe influenced both its Cold War approach to defense and its post-Cold War space aspirations.

Early space-related activities in Chile date back to 1959, when the Center for Space Studies was created at the Universidad de Chile, and

Chile entered into an agreement with NASA. A satellite receiving station was built near Santiago for the Apollo program. This began a decades-long history of Chilean space cooperation with the American space agency. In the 1985 Mataveri Agreement, Chile's Easter Island became an emergency landing site for US space shuttles. After the restoration of democracy in 1989, Chile's official space program took off, but like that of Chile's neighbor, Argentina, it had a military progenitor.

In May 1991, the Centro de Estudios Aeronáuticos y del Espacio (Aeronautic and Space Studies Center—CEADE) was founded as an arm of the Chilean air force. Its mission was to study the political, economic and social variables important to the aerospace industry and to national defense. Chile's first two satellites were sponsored by the Chilean Air Force (FACH). The *Fasat-Alfa* (Spanish acronym for "Chilean Air Force Satellite-Alfa") was launched from the Plesetsk Cosmodrome on 31 August 1995 as a communications, ozone-monitoring, and communications satellite through a technology transfer program with Surrey Satellite Technology Limited (SSTL), a British company that specializes in a standardized series of satellites that can be easily adapted for remote-sensing, scientific, and navigation purposes. As with many satellites of other EMSAs, *Fasat-Alfa* existed in part to give Chilean engineers hands-on practice in satellite construction and control, from a brand-new control station erected in Santiago. However, *Fasat-Alfa* was lost when it failed to separate properly from its launcher. The replacement *Fasat-Bravo* (also built by Surrey) was launched on 10 July 1998 from Baikonur and also focused on remote-sensing platforms, with specialized cameras for ozone studies. Both were owned and operated by the FACH. With its first success in hand, the FACH pushed forward a motion to create a true space agency.

Based upon Decree No. 338 from the Defense Ministry, the Agencia Chilena Espacial (Chilean Space Agency—ACE) was founded by President Ricardo Lagos in 2001 as an advisory commission to the Office of the President.[64] The stated mission of the ACE is to utilize space activities to monitor and improve a host of economic activities in forestry, fishing, and mining as well as geological research and natural disaster monitoring. In addition, the ACE is charged with promoting international cooperation in space activities. Noteworthy in the creation of the ACE is that the leadership in space activities has been taken from the FACH and made civilian in the person of the director of the National Commission for Scientific and Technological Research (CONICYT).

The first objective of this new space agency was to put into orbit an observation satellite to support the aforementioned socioeconomic activities. This satellite effort is the advanced US$72-million Sistema Satelital para la Observación de la Tierra (Satellite System for Earth Observation—SSOT), which will be an Earth observation satellite built by the European EADS consortium, but again with the involvement of Chilean scientists and engineers. The SSOT will be launched by the ESA aboard a Soyuz launcher from French Guiana in 2012, and will provide Chile with extremely high quality images for use in cartography, urban planning, agricultural management, mineral and oil resources, and natural disaster response. Nevertheless, as it is a satellite of the FACH, the Chilean military will also have privileged access to all data. Another example of Chile's satellite construction progress is the CESAR-1 (Chile Satellite for Amateur Radio), which is being built for amateur radio communications by Chile's Universidad de La Frontera. At the bargain price of US$575,000 (US$350,000 from private contributions) and measuring only 20×20 centimeters, CESAR-1 represents the newest wave of microsatellite technology that many smaller EMSAs are embracing because of its low entrance costs.[65]

In addition to its forays into satellite ownership, Chile has distinguished itself as a space science leader in Latin America. Early projects included joint ventures with NASA and the ESA, which have included the Paranal Observatory in the Atacama Desert, which houses the appropriately named Very Large Telescope (VLT), a joint venture with the European Space Agency. Chile spends half a billion dollars annually on space investments and satellite services.[66] Its strong investment in and dedication to space-related activities has promoted Chile's aspiration to becoming a regional leader in space cooperation. In 2002, the ACE proposed the creation of a Latin American space agency to Argentina, Brazil, and Mexico, in hopes of mirroring the multinational European Space Agency.

Africa's celestial poverty of wealth

Besides South Africa, the only other African states that have established formal space policies are Algeria, Egypt, Tunisia, and Nigeria (see Table 5.2). In the Egyptian case, space activities, like in Israel, have been largely dictated by security concerns. For a number of other African states, however, the socioeconomic benefits have taken center stage in the past couple of decades, despite the relatively high cost. Space technology has been seen by both the public and private sectors in Africa as a potential contributing factor to these states' long-term development, despite the plethora of obstacles to their economic, social, and political development. However, a deficit of modern technology is not among them. Even in the relatively poor countries of Africa, space-based technologies increasingly play a part in socioeconomic development schemes.

Among space-borne technologies, global positioning systems (GPS) and mobile phones are progressively becoming standard tools to help isolated communities in these countries to become more integrated into national and global economies. Topographic maps or images of villages, watersheds, or entire countries put invaluable tools in the hands of local officials and developers to determine property status and land use rights, and to utilize resources at a hitherto unimaginable level of sophistication.[67]

Simply by using these satellite-provided data, countries could free up hundreds of billions of dollars worth of potential economic development, since competing land-rights claims in the absence of clear and consistent land title records have for a long time been a major sticking point in the full utilization of land in the developing world. Compared to hard-wired systems, mobile phone systems are infinitely cheaper to set up and require practically no training for their use, and the availability of telecommunications satellites from a variety of providers eases their implementation. For those few African countries with relatively more financial resources, which are typically the petro-states, the decision to implement space-based resources has come to fruition in recent decades.

Egypt: the return of Ra

With a missile industry second only to Israel's in the region, and with occasional North Korean and Chinese assistance, Egypt has for the past 50 years sought to improve its missile capabilities. Egypt's initial rocket efforts began in the 1960s with German assistance in the construction of Factory-333 in Heliopolis. Three rockets were being developed: *al-Zafar* (375-kilometer range), *al-Kahar* (600 kilometers), and *al-Raid* (1,000 kilometers); however, the project ground to a halt in 1966 when the West German government withdrew cooperation. In October 1984, Egypt signed an agreement with Argentina to begin the development of the *Cóndor* missile project. During the 1980s and 1990s, Egypt extended the range of its *Scud* missile with North Korean and Chinese help, testing its first indigenously designed and built missile, *Amun-2*, in 1990.

Table 5.2 Third tier space actors of the Middle East and Africa

Country	Space agency	Principal activities
Egypt	Egyptian Space Science and Technology Research Council	• space applications • satellite control • Bilateral/multilateral cooperation
Nigeria	National Space Research and Development Agency (NASRDA)	• space applications • satellite control • bilateral/multilateral cooperation
Algeria	L'Agence Spatiale Algérienne (ASAL)	• space applications • satellite control • bilateral/multilateral cooperation
Tunisia	National Commission for Outer Space Affairs (NCOSA)	• space applications • satellite control • bilateral cooperation

Probably because of its position as the recipient of US$1.8 billion in annual aid (as of 2009) from the United States, Egypt has not chosen to pursue the next step of independent launch capability but has opted, like most developing world EMSAs, to purchase satellites from established companies. Egypt's first satellite was *Nilesat 101*, a geosynchronous communications satellite, built by Astrium and launched from Kourou in 1998. The companion *Nilesat 102* (also built by Astrium) joined its predecessor in geosynchronous orbit in 2000. Through its National Authority for Remote Sensing and Space Sciences, Egypt's more ambitious undertaking was its decision to co-manufacture its first remote-sensing satellite, *Egypt-Sat-1*, with the Ukrainian Yuzhnoye Design Bureau for purpose of technology transfer. The 100-kilogram *EgyptSat-1* was launched from Kazakhstan on 17 April 2007. Though officially classified as a scientific satellite, *Egypt-Sat-1* provided one of Israel's neighbors, for the first time, with comparable intelligence-gathering capabilities (it failed, however, in July 2010).

The successor *EgyptSat-2* was slated for launch into LEO in 2013 and the remote-sensing *Desert-Sat* for 2017, but the Arab Spring revolt of 2011 that ousted President Mubarak put an indefinite hold on both. Once they finally reach orbit, both satellites will provide imagery for land-use planning and socioeconomic development, though the latter, as the name implies, will be focused on water and potentially petroleum deposits in Egypt's vast deserts. Two satellite-receiving stations are currently in place in Cairo and Aswan.[68]

Nigeria: turning oil into space

By contrast, Nigeria's space policy has been entirely motivated by developmental concerns and entirely absent the strategic defense concerns historically present in South Africa and Egypt. As Africa's most populous country (155 million) and possessing a land area twice the size of California, Nigeria is rife with maladies all too common in the sub-Saharan region, such as disease, poverty, corruption, and malnutrition. Despite its petroleum industry generating over US$60 billion in annual revenues, Nigeria is one of the poorest countries on Earth with 70 percent of its population living in poverty. As of 2010, Nigeria's per capita GDP was only US$2,500 and its national annual budget just over US$13 billion, with an estimated US$3.3 billion in foreign debt in 2008. Outside the country's largest cities, most Nigerians do not have running water or electricity, 30 percent of the population is illiterate, and most roads are still dirt paths. But, despite these daunting developmental obstacles, the Nigerian government sees its nascent space program as a wise and necessary investment in the country's continued social and economic development.

The Nigerian National Remote Sensing Center became operational in 1996 and in 1998 a committee was selected to draft a space science and technology policy. The National Space Research and Development Agency

(NASRDA) was founded in 1998 and is part of the Federal Ministry of Science and Technology. The initial budget was US$93 million. The following year, flush with cash from Nigeria's lucrative petroleum exports, which make up 95 percent of the country's foreign exchange earnings, the Nigerian government approved the "Nigerian Space Policy." This policy foresees a national space program that would build capacity in science and engineering as well as socioeconomic development, declaring that "there is no nation that can call itself developed in the twenty-first century that does not have indigenous critical mass of trained space scientists and engineers who contribute actively to the solution of the nation's problems."[69]

The stated mission of NASRDA reflects the socioeconomic strategic focus of the country's investment in space: "to build indigenous competence in developing, designing, and building appropriate hard and software in space technology as an essential tool for its [Nigeria's] socio-economic development and enhancement of the quality of life of its people."[70] The mission is specifically targeted toward addressing some of Nigeria's most pressing development problems in communications, telemedicine, and remote sensing for agriculture.[71] Concurrently, the mission calls for using the space program to increase Nigeria's scientific, industrial, and academic sectors to further the country's self-reliance. The NASRDA is subdivided into six geographically and mandate-distinct centers: (1) Centre for Satellite Technology Development (located in Abuja); (2) Centre for Space Transport and Propulsion (in Epe); (3) Centre for Basic Space Science and Astronomy (in Nsukka); (4) Centre for Space Science Technology Education (in Ile-Ife); (5) National Centre for Remote Sensing; (6) Centre for Geodesy and Geodynamics (in Toro). To begin the process, NASRDA's first objective of launching a satellite was announced in 2001.

Like most other EMSAs that have established a national space policy, Nigeria began its space journey with both a plan and a foreign space technology purchase. In 2000, Nigeria signed a contract with Surrey Limited for a microsatellite called *NigeriaSat-1*. The US$13-million satellite's mission was communications and disaster-monitoring, the latter being Nigeria's contribution to the international Disaster Monitoring Constellation (DMC) project. The satellite represented a significant financial and development investment for a poor country but was part of Nigeria's long-term space plan. Fifteen Nigerian engineers were trained in Britain and were involved in the design and production of the satellite as part of a technology transfer program co-sponsored by the University of Surrey.[72] The 98-kilogram *NigeriaSat-1* was launched from Plesetsk in September 2003 and put into a sun-synchronous orbit at an altitude of 686 kilometers. The investment seems to have paid off well for Nigeria, both in terms of building knowledge and technology infrastructure but also as a budgetary subsidy. In 2006 a United Arab Emirates–based company signed a deal with NASRDA worth US$250 million for bandwidth access to *NigeriaSat-1* and a further US$140 million annually is expected from direct sales and servicing of the transponder from the satellite.[73]

A follow-up satellite, *NigeriaSat-2* (also built by Surrey) was launched aboard a Russian Dnepr-1 launch vehicle from the Yasny Launch Base in southern Russia near the Kazakh border on 17 August 2011. This latest Nigerian satellite is reputed to be among the most advanced micro-satellites yet developed. The 2.5-meter high-resolution, remote-sensing satellite will provide imaging for applications in mapping, water resources management, agricultural land use, population estimation, health hazard monitoring, and disaster mitigation and management. The satellite will also aid in the creation of Nigeria's autonomous geographical information system through high resolution geospatial data and images. *NigeriaSat-2* will enable Nigeria to join the second-generation Disaster Monitoring Constellation (DMC), a system of satellites in a consortium including Algeria, Turkey, United Kingdom, Spain, and China. This newest satellite provides Nigeria with remote-sensing capabilities, making it the only such enabled country in central Africa and providing Nigeria with a considerable commercial advantage in the region in resource mapping. Launched along with *NigeriaSat-2* was the 100-kilogram *NigeriaSat-X*, an Earth observation satellite with a 22-meter multispectral imaging system with a 600-kilometer swath.

These two satellites represent just the beginning of Nigeria's ambitious long-term space plans, which set some lofty milestones indeed for a developing country: by 2015, to produce a Nigerian astronaut; between 2018 and 2030, to use its advantageous equatorial location to build a spaceport and become the first African country to autonomously launch its own satellite; and by 2030, to land a Nigerian astronaut on the moon.[74] Whatever the feasibility of these grand aspirations, they reveal the important place that space policy has assumed among even the poorer countries of the world as a means to achieve hitherto unthinkable developmental and prestige opportunities.

Algeria: colonial space legacy

For Algeria, space activities have been a part of its history since its birth as an independent country. Shortly after the country gained independence from France in 1962, Hammaguir in Algeria was used as the site of France's sounding rocket tests as well as France's first (and the world's third) satellite launch in 1965. Algeria was also the site of France's nuclear weapons tests until 1966. Algeria has come full circle to again embrace space activities.

In 2002, Algeria created the L'Agence Spatiale Algérienne (The Algerian Space Agency—ASAL) with the official policy goal of eventually reaching satellite-production independence. Algeria's National Space Technology Centre purchased its entry-level US$15 million *Alsat-1* satellite from Surrey; a joint British-Algerian team worked on it in a technology and knowledge transfer program, giving Algerian engineers and scientists valuable satellite construction experience. A mission control ground station was also constructed in Arzew, Algeria, 30 kilometers east of Oran, Algeria's second-largest city.

On 28 November 2002 *Alsat-1* was launched aboard *Kosmos 3-M* into a sun-synchronous orbit from the Plesetsk Cosmodrome in Northern Russia. Algeria's first satellite became part of five satellites in the Disaster Monitoring Constellation operated for the Algerian, Nigerian, Turkish, British, Vietnamese, Thai, and Chinese governments by DMC International, a Surrey subsidiary that provides Earth-imaging for disaster relief operations. The follow-up US$17-million *Alsat-2A* was sent aloft on 12 July 2010 from India's Sriharikota Launch Center to join its predecessor.[75] More than 20 Algerian engineers received experience in satellite construction, system operations, and satellite integration, this time completed by the European Aeronautical Defence and Space consortium (EADS) to propel Algeria further toward gaining autonomy for its space program. The *Alsat-2A* possesses a spatial resolution of 2.5 meters and downloads its data to Algerian controllers in Algeria rather than to foreign controllers.

In 2006, the ASAL adopted a 15-year, 10-satellite program called "Space Programme 2020," which included *Alsat-2B*, a remote-sensing satellite for taking high-resolution images for topographic studies, agricultural planning, natural disaster monitoring, mineral and petroleum development, and the detection of potentially devastating locust swarms.[76] The *Alsat-2B* was launched from Sriharikota on 12 July 2010, and was the Algerian first satellite assembled at the domestic satellite development center in Oran (though built by EADS Astrium). In addition, Algeria has furthered its commitment to its space activities by signing several cooperation agreements with the space agencies of South Africa, Argentina, Russia, France, United Kingdom, and the United States. In January 2009, Algeria entered into an agreement with the United Nations Platform for Space-based Information for Disaster Management and Emergency Response (UN-SPIDER) to facilitate the development and use of space-based imagery for the prevention and management of natural disasters.

Tunisia: steady and skyward

To the east of Algeria, the small state (population 10 million) of Tunisia has similarly sought to utilize satellite technology for national development. For a long time, Tunisia has had one of Africa's best performing economies and the second highest GDP per capita in North Africa and the Middle East. Because of its stability (until the 2011 Arab Spring uprising) and strong trade ties with the European Union (comprising about three-quarters of Tunisia's imports and exports), Tunisia has been notable for attracting investment from large high-tech companies such as Airbus and

Hewett-Packard. The country's history of stability helps to explain its natural interest in investing in space for its development. Conspicuously, Tunisia is one of only 20 countries to have established national legislation governing space activities, and it participates in a number of international space-related projects for the Maghreb countries.[77]

In 2001, Tunisia furthered cooperation with fellow Maghreb states to establish a communications network that was envisioned to utilize satellite communications and data-sharing to further develop the region's educational systems. At a 2005 meeting of the UN Summit on the Information Society in Tunis, Tunisia's long-time president, Zine El Abidine Ben Ali, praised satellite-based technology as a way "to reduce disparities between peoples, and ensure a balanced, safe and equitable information society."[78]

Coordinating the country's use of space-based technologies is the Tunisian National Commission for Outer Space Affairs (NCOSA), which was established in 1984 to oversee the activities of the various ministries' work in space technologies. In addition, NCOSA seeks to generate greater domestic understanding in Tunisia of the benefits derived from space technologies. The principal space-based activities planned are telecommunications, Earth observation, and remote sensing. For example, the Tunisian Ministry of Agriculture has made extensive use of satellite data for cartography, monitoring and evaluating natural resources. A pilot remote-sensing project had been commenced in 1975 with French cooperation. This collaboration, known as Arid Zones of Tunisia (ARZOTU), utilized satellite images of the country's arid environments to study water resources, land use, erosion mapping, and agricultural production.[79] Because of its experience, Tunisia has also played a key role in the Arab world in the promotion and growth of satellite technology.

The National Center for Remote Sensing was founded in 1988, and since 1990 the Tunisian capital of Tunis has housed the main office of the North African States Regional Center for Remote Sensing, whose members include Algeria, Egypt, Libya, Mauritania, Morocco, Sudan, and Tunisia. Perhaps most noteworthy is Tunisia's role as an integral player in the operation of a secondary satellite control center for the Arab Satellite Communications Organization (Arabsat), founded in 1976 by 21 members of the Arab League and serving 164 million viewers in over 80 countries.[80] Concurrently, Tunisia has been developing a compendium of Arabic-language terminology appropriate for use in satellite communications systems. To fulfill its current and future ambitions, Tunisia has devoted 1.25 percent of its annual federal budget to scientific research, and in 2009 Tunisia became a member of the International Astronautical Federation (IAF) for its work in remote sensing.[81]

Asia: font of the next space race?

It now popular among international observers to declare that the twenty-first century will be the "Asian century," because of the vast continent's positive demographic and economic growth trends. In addition to the still-economically successful Japan, the region will hold the largest proportion of the world population and two of the world's largest economic powers— China and India—by 2050. For these two countries (as discussed in Chapter 3), seeking prestige, socioeconomic benefits, and national security through space programs is par for the course. But apart from these largest Asian space actors, many more Asian states are emerging as regional economic powers in recent decades as part of the Asian Tiger phenomenon (see Table 5.3). In line with their newfound economic prosperity, many of these states have enthusiastically adopted space policies that make the assumption that continued development will rely on space assets.

Table 5.3 Third tier space actors of Asia

Country	Space agency	Principal activities
North Korea	Korean Committee of Space Technology	• launcher construction • satellite construction • satellite control
Kazakhstan	National Space Agency of the Republic of Kazakhstan (KazCosmos)	• space applications • satellite control • bilateral cooperation
Azerbaijan	Azerbaijani National Aerospace Agency (ANASA)	• space applications • satellite control
Vietnam	Vietnam Space Technology Institute (VSTI)	• space applications • satellite control • bilateral cooperation
Indonesia	National Institute of Aeronautics and Space	• sounding rockets • space applications • satellite control • bilateral cooperation
Bangladesh	Space Research and Remote Sensing Organization (SPARRSO)	• space applications • satellite control
Pakistan	Pakistan Space and Upper Atmosphere Research Commission (SUPARCO)	• sounding rockets • space applications • satellite control
Malaysia	National Space Agency of Malaysia (ANGKASA)	• space applications • satellite control
Taiwan	National Space Organization	• sounding rockets • space applications • satellite control
Philippines	Aeronautics and Space Agency	• space applications • satellite control
Thailand	Geo-Informatics and Space Technology Development Agency	• space applications • satellite control

With the exception of odd-man-out North Korea, these newest Asian EMSAs have implemented space policies that emphasize areas from pure science, to remote sensing for socioeconomic purposes, to telecommunications. In fact, three-quarters of Asian countries' first satellites have been for telecommunications, suggesting the strong developmental priorities of their space policies. Moreover, these first efforts, like in all other cases, provide test beds to improve indigenous space technologies. But while the focus of these programs is purely scientific and/or socioeconomic, these countries' space policies nonetheless are still in line with the general thrust of the argument of this book: that space programs primarily service states' national interest, which is at the heart of the realist paradigm that has driven space developments over 70 years. This is true despite that fact that the thrust of the space investment in these third tier countries is hoped to yield benefits in greater education, economic growth, and/or environmental security. Since these and many other developmental areas enhance a country's well-being and security, the activities still fall under a more inclusive definition of national security. In other words, whether the space technology produces more readers, more vaccines, or more missiles, the ultimate goal is making the state stronger and wealthier, thus providing the socioeconomic advantages that preserve national sovereignty.

North Korea's enigmatic space gambit

Whenever the subject of North Korea arises, almost without fail, discussions of its secrecy and massive security state ensue, and for good reason. Since the tenuous 1953 ceasefire that put the Korean War on standby, North Korea has put together the most comprehensive Orwellian security state in modern history, arguably more comprehensive than the USSR under Stalin. In line with this security mania, North Korea has aggressively pursued the development of the missile, nuclear, and space triad with varying levels of success. Given the monolithic structure of its state apparatus, North Korea has no separate space agency because all space activities fall under the manifest rubric of national security, though officially, the state-run Korean Committee of Space Technology coordinates space activities.

With an eye toward thwarting Western governments' attempts to isolate it, North Korea has doggedly pursued a policy of self-reliance (*chuch'e*) in missile technology, and has become one of the largest exporters of ballistic missile technology in the developing world, actively assisting a wide and varied collection of other states to develop their own systems. This process began in 1960 through an agreement reached with the Soviet Union to help modernize the North Korean military arsenal, including the introduction of its first missile system, the SA-2, which was deployed in early 1963. But North Korea's support of China during the Sino-Soviet split (1960–89) meant the end of Soviet help and the beginning of Chinese-led development. Under China's guidance, North Korea took its own tentative steps toward building Type 63 multiple rocket launchers in the early 1960s. The Hamhung Military Academy was founded in 1965 to provide a dedicated program in missile training, and two renewed missile assistance pacts with the Soviet Union were signed in 1965 and 1967.[82] In the late 1960s, Soviet-built *Frog-7* tactical rockets were added to the country's arsenal, and in the mid-1970s *Scud-B* missiles (renamed *Hwasong* in North Korea) were purchased from Egypt .

The path from missile development to a space program in North Korea, like so much else in the country, was unabashedly imbued with realist security concerns. The space program was set in motion by the establishment of a ballistic missile program in 1975 in response to a similar program in South Korea, whose own missile program, in turn, had been spurred by the North Korean Frog and the withdrawal of some US forces from South Korea.[83] Probably spurred by an again-renewed friendship with China, North Korea procured the Chinese DF-61 ballistic missile, which provided a starting point for North Korea's budding indigenous ballistic missile force.[84]

By 1984, North Korea had produced and flight-tested an indigenous version of the *Scud-B*, and in the next year the country signed an agreement with Iran for financial assistance for missile development and production, for which North Korea would sell Iran discounted missiles.[85] Despite reverse-engineering these acquisitions, further development of an indigenous ICBM was hampered by a significant deficit of trained engineers, but again Egypt filled the void for a fellow authoritarian developing state, providing an Egyptian R-17E copy of a Soviet *Scud*. This technical cooperation was mutually beneficial to the long-term security and self-sufficiency goals of both countries, at least from the hegemonic power perspective of the Cold War.[86] The R-17E was hurriedly reverse-engineered and renamed *Hwasong 5*, which converted North Korea into a potential exporter of missile technology.[87] Between 1977 and 1987, North Korea earned almost US$4 billion from arms transfers and missile technology to over thirty countries in Africa, the Middle East, and Central America.[88] The United States Arms Control and Disarmament Agency estimated that North Korea's new role as exporter also had the effect of providing a noteworthy and needed source of hard currency for Pyongyang as well as affording it an opportunity to see its missiles tested on the battlefield during the Iran–Iraq War, without the difficulty, tensions, and scrutiny that test flights over the Korean peninsula would generate vis-à-vis the United States and South Korea.

The 1990s saw the North Korean program closer to actual space flight. In 1993 Kim Il-Sung officially declared that space-related developments were now desirable, thus publicly couching the missile program within the framework of an internationally protected space program via the Outer Space Treaty.[89] But after Kim Jong Il's assumption of power in 1998, North

Korea defiantly adopted two key policies that would reinforce the overt security nature of its space program. The first was the lofty policy of *kangsŏngdaeguk* (strong and powerful country). This policy aspires to develop defensive capabilities that guarantee national political independence as well as a self-reliant economy, and declares that "the nation can become prosperous only when the barrel of the gun is strong."[90] The second policy was a directive called *sŏn'gunjŏnghi'i* (military-first policy), which prioritized all resources for military use over civilian needs.

In August 1998 North Korea test-launched its home-grown *Taepodong 1*, a 25-meter, 21-ton, medium-range ballistic missile (called *Paektusan-1* in North Korea after the alleged birthplace of Kim Jung Il). Originally intended to carry a 1,000-kilogram warhead, it was fired from the launch facility at Musudan-Ri, in the northeastern corner of the country, sending the missile on a trajectory that carried it menacingly over the main Japanese island of Honshu. The *Taepodong-1* emerged as the country's space launch vehicle, boasting either two- or three-stage variants and capable of hoisting up to 1,000 kilograms into orbit.[91] The following month, the North Korean Central News Agency reported to the world the successful launch of the country's first indigenous satellite, *Kwangmyŏngsŏng-1*, which coincided with the country's fiftieth anniversary. It was also an act that can be attributed to a Korean peninsula mini-space race (South Korea had successfully launched its first satellite in 1992). Despite the North Korean government's claims of success, there was no independent verification by any developed space actor of a successful orbital placement, though a debris trail was noted along the satellite's intended trajectory, suggesting it had broken up before reaching orbit.[92]

Amid fears that North Korea's missile testing might in fact be a façade for testing technology that would be used for a future nuclear-tipped ICBM, the UN Security Council passed Resolution 1718, a measure that effectively banned further missile or nuclear testing by the reclusive state. Under pressure, North Korea agreed to a missile test moratorium in September 1999. Nonetheless, design work and static testing of the successor *Taepodong-2* continued unabated. On 5 July 2006, North Korea abandoned its self-imposed suspension and fired off seven test missiles that possessed double the range of their predecessors. The first failed only 40 seconds into flight, but nevertheless impressed observers with its improved range, which suggested North Korea was ready for more ambitious projects.[93]

By 2009, North Korea's MNS triad was reaching realization. Perhaps to legitimize its declared intention of the use of space for peaceful purposes, North Korea signed the Outer Space Treaty in March 2009 and informed the UN International Civil Aviation Organization (ICAO) of its next launch. On 4 April 2009, despite the UN resolution and warnings from Japan that it might shoot down the missile, North Korea defiantly launched a three-stage Taepodong-2, which carried the country's second satellite, *Kwangmyŏngsŏng-2* (which had been completed with the assistance of a number of Iranian engineers).[94] Again, a North Korean missile crossed Japanese airspace, contributing to a long-running campaign of psychological warfare directed at Japan, South Korea, and the United States. And again, although North Korea again claimed success in placing a satellite into orbit, Russia, the United States, and other observers reported the failure of the third stage as well as the missile's destruction in the Pacific Ocean.[95]

Following the launch, the UN Security Council issued a statement that branded the launch a violation of Resolution 1718. In response, North Korea defended its program as being protected under Article I of the Outer Space Treaty and updated the launcher's name to *Unha-2* to differentiate its public purpose as a space vehicle. Despite the failure, the launcher had flown over 3,200 kilometers, doubling the previous effort and suggesting that North Korea is closer to truly joining the space club. The very next month, on 25 May 2009, North Korea conducted a second underground test of an indigenous one-kiloton nuclear weapon, which was verified by Western seismic readings of a 4.7 on the Richter scale, located 375 kilometers northeast of Pyongyang.[96] This test was the culmination of a nuclear program that had been decades in development at North Korea's nuclear research center in Yongbyon, though according to an old Soviet KGB report, no testing had been conducted out of concern for detection and the certain international ramifications.[97] It is estimated that North Korea has completed between three and nine low-yield, Nagasaki-type weapons (between 10 and 20 kt).

Throughout most of the past decade, North Korea has been constructing a new launch facility, this time on the west coast of the country. Called Tongch'ang-dong, the facility is located about 110 kilometers northwest of Pyongyang and 47 kilometers from the Chinese border, making it more difficult for US drones to observe directly. The facility is thought to be much more sophisticated than the Musudan-ri complex in that it possesses a number of improvements that will make it a more complete launcher test, research, and development center. It is a much larger facility and possesses a longer launch vehicle gantry umbilical tower, which will allow for future expansion beyond the current *Taepodong-2*.[98] Moreover, its location will allow for southern launches that will not overfly Japan or South Korea. In 2009 North Korea announced even more ambitious future space projects including its own manned space flights and development of a manned partially reusable launch vehicle.[99]

In the end, North Korea's burgeoning space program has two distinct purposes. First, the program serves as a bulwark in its strategic position vis-à-vis its southern cousin (and, by extension, the US) and, second, it is a powerful propaganda tool that has been exploited by Kim Jong Il and his successor to bolster national pride, despite the high opportunity costs of developing the technology. In other words, North Korea is taking pages from the Cold War space playbooks of the Soviet Union and the United

States to bank on prestige gains along with real technology gains in its nascent space program as a way to build domestic political capital and maintain the status quo on the Korean Peninsula.

Central Asia: inheritors of a space legacy

For many of the former Soviet republics in Central Asia, space programs were a legacy of the communist period as the USSR utilized the region's geography, industry, and manpower during the space race to fulfill Soviet state policy. But the dissolution of the Soviet Union in 1991 meant that for two of the former republics, their former position was a stepping stone toward viable programs of national development via space technology.

With only 15 million inhabitants, Kazakhstan's small population belies its geography as the largest of the former Soviet republics, the world's largest land-locked country, and the owner of one of the world's most important space ports. In the 1950s, an existing missile test center in south-central Kazakhstan was chosen by Soviet authorities as the rechristened Baikonur Cosmodrome, becoming the Soviet Union's primary space launch facility. Located on more than 100 kilometers of rolling steppe, the featureless flat plains and the region's relatively sparse population were attractive characteristics for the budding space program for reasons of both downrange safety and security. Typical of Soviet projects of the day, a massive public works project was undertaken that built hundreds of kilometers of new roads and a railroad line as well as a new town to support the workers. It became, and still is, the world's largest space launch complex, hosting over a dozen launch pads. Many of the Soviet space firsts, such as the *Sputnik* satellite and Yuri Gagarin's orbital flight, were launched from Baikonur.

Following the breakup of the Soviet Union, Kazakhstan made a number of hard choices regarding the future of its inherited space and missile assets. In 1992 Kazakhstan ratified the START I treaty, and the following year all intermediate-range missiles were removed by Russia. Likewise, in 1994 Kazakhstan joined the Non-proliferation Treaty and the process of removing Russian ICBMs and nuclear weapons from Kazakh territory began. But, because Russia's largest operational rocket, the *Proton*, still needed to be launched from Baikonur, Russia entered into a 20-year lease agreement in 1994 with Kazakhstan to continue using Baikonur for an annual lease of US$115 million. Those terms were eventually extended to 2050. Russia also leases many other sites in Kazakhstan for a variety of aerospace activities, including a space radar facility in Gulshad, on the shores of Lake Balqash, which monitors satellite and ballistic missile activity over central Asia, and a weapons testing range at Aktobe.[100]

However, the changing fortunes of Kazakhstan's space ambitions are possible because of the country's windfall profits from petroleum and other natural resource extraction (e.g., the world's largest production of uranium), which produced an average real GDP growth of 10 percent from 2000 to 2007. With such strong wind in its economic sails, the Kazakh government has become more interested in sharpening their country's international image and financial status. The logical decision was to use its bequeathed Soviet space assets to begin to pursue an independent space program, which was part of President Nursultan Nazarbayev's vision of his country exiting its "Third World" status.[101]

In 2004, Nazarbayev unveiled a national space policy, which envisaged a national space program that would function as a strategic instrument contributing to the country's economic and security needs. One of the first goals of the program was to become at least somewhat independent of Moscow, beginning by gradually replacing Russian engineers at Baikonur with Kazakh engineers.[102] A further step was to construct by 2017 the new US$1.3-billion Baiterek Space Launch Complex, an environmentally friendly complex at Baikonur that will house new processing and launch facilities to support launches of Russia's Angara launch vehicle (the long-used Soyuz launch vehicle utilizes a poisonous fuel mixture). But one of the likely motivations behind the sudden urgency of Kazakhstan's rush into space activities is the desire to be able to utilize Baikonur itself in the wake of Russia's scheduled partial departure. Russia's new Vostochny Cosmodrome in Omur Oblast in the Russian Far East is scheduled for completion by 2018, and will replace Baikonur as Russia's launch site for its *Proton* launcher and also become its primary launch site for human space flight.[103]

For its first space project, Kazakhstan contracted Russia's Khrunichev Design Center to build Kazakhstan's first national satellite, the US$65-million communications satellite *KazSat-1*, which left Baikonur on 6 June 2006 atop a Proton-K launcher. The follow-up, Russian-built *KazSat-2* communications and data platform was launched in July 2011. The planned *KazSat-3* communications satellite as well as a yet-unnamed remote-sensing satellite are scheduled for launch in 2014. In March 2007 Nazarbayev created the National Space Agency of the Republic of Kazakhstan (Kazcosmos), though the national joint-stock company Kazakhstan Gharysh Sapary (KGS). Kazcosmos is charged with the actual management of the space program via coordination of national industries and implementation of national space policy. While carving out a national character for its space program, Kazakhstan has continued its long-standing cooperative arrangements with Russia as well as other former-Soviet republics. In "pure" space science, Kazakhstan has joined a consortium including Russia, Germany, and Spain to build and operate the World Space Observatory UltraViolet (WSO-UV), a space telescope that will conduct observations in the ultraviolet domain. The launch of the WSO-UV is planned from Baikonur in 2015.

A 2009 white paper called "Development of Space Activities in the Republic of Kazakhstan for 2009–2020" outlined Kazakhstan's mid- and long-term space program objectives. These objectives include the creation of a national remote-sensing system with two Atrium-built remote-sensing satellites for cartography, mineral and energy exploration, and natural disaster mitigation as well as a GPS navigation satellite, both scheduled for launch by 2012.[104] Also scheduled for completion by 2012 is the construction in Astana of the Assembly and Testing Complex of the Engineering and Design Office of Space Technology, which is being carried out by Atrium-EADS.

Despite the manifest diversification and autonomy-building measures, traditional security concerns have not been ignored within Kazakh space policy. In delineating his space agency's ambitions through 2020, President Musabayev stated that Kazakhstan has planned to set up a space reconnaissance and flight-control system, which will provide for the positioning of troops and high-precision weapons system. There are also plans to take part in a collective [within the CIS] missile attack warning system, which is integrated into the space monitoring system of state borders and others.[105]

Another former Soviet republic to move robustly toward implementing a space policy is Azerbaijan, which like Kazakhstan has enjoyed recent exploitation of oil and natural gas wealth, though poverty persists for around 40 percent of its eight million citizens. Revenues of the state-owned petroleum corporation SOCAR—the 68th largest company in the world—are expected to produce US$160 billion over the next 15 years, thus enabling substantial investment in space assets.[106] During the Soviet period, the Scientific and Industrial Association for Space Research was established in Baku in 1975 as part of the infrastructure of the Soviet space industry.

Azerbaijani specialists contributed significantly to the USSR's space science. Especially noteworthy was Kerim Aliyevich Kerimov, an Azerbaijani rocket specialist who was one of the founders of the Soviet space industry, and a leading architect of many Soviet space achievements from *Sputnik* to the space station *Mir*. The Azerbaijani National Aerospace Agency (ANASA) was originally established to coordinate the production of equipment for the Soviet-era space program. Following the fall of the Soviet Union, Azerbaijan's national space activities came under the direction of the ANASA, which joined the International Astronautical Federation in 2003.

In September 2006 ANASA was put under the jurisdiction of the country's new state-owned Ministry of Defense Industry, which had been established to nationalize the production of military hardware. This organizational place for ANASA strongly suggests the policy importance that space has acquired for Azerbaijan in recent years and possibly the long-range focus of the country's space policy. Azerbaijan's first satellite, a US$202-million communications platform called *AzerSat-1*, is being built by Orbital Sciences Corporation in Virginia and is scheduled for launch by

Arianespace from the Guiana Space Center in mid-2012.[107] Only 20 percent of the bandwidth will be utilized by Azerbaijan, so the satellite's broad coverage of much of Central Asia and Eastern Europe will offer the opportunity to lease satellite television and radio broadcasting services to Asia and Europe and will provide communications coverage to the whole of Azerbaijan's mountainous territory.

The Azerbaijani government views the satellite communications sector as a way to diversify the country's dependence on petroleum exports. In 2009, President Ilham Aliyev announced a policy entitled "State Program on Establishment and Development of Space Industry in Azerbaijan," which makes space activities an official priority.[108] The policy plainly declares that space activities have "a great importance from the standpoint of providing national security" and is a step toward "the elimination of dependence on data exchange from foreign countries."[109] Of primary interest to Azerbaijan in future satellite projects is to expand its capabilities in remote sensing, especially in using the data to develop tectonic maps which can, indirectly, offer indications of probable oil and gas fields, especially in the vital Caspian Sea region.

Cooperative arrangements have been carried out as well. The ANASA has worked with the UN FAO in a project that used GIS technology to produce land-use maps for all Azerbaijan, which will be used for planning agriculture production and environmental mapping of the more arid areas proximate to the Caspian Sea.[110] Azerbaijan is set to embark upon the construction of assembly and production of satellite receiving and transmitting stations as well as the construction of a satellite assembly facility by 2013.

Vietnam: the other East Asian EMSA

Vietnam was among a number of states to contribute to InterCosmos, a Soviet space program with specific designs to strengthen political as well as scientific relationships among both Warsaw Pact and non-aligned countries during the Cold War. In 1980 Maj. Gen. Pham Tuanhas became the first Vietnamese cosmonaut and the first Asian in space, on board Soyuz 37, thus providing both the USSR and Vietnam with a space prestige event. In the waning days of the Cold War, Vietnam adopted free market reforms in the mid-1980s, following the Chinese model of economic liberalization without political liberalization. From 1990 to 2007, economic growth averaged 7.5 percent, making Vietnam one of the world's fastest growing economies.

Now flush with cash, the Vietnamese government has again embraced a space program, but this time for its socioeconomic benefits rather than its military accruements. Since 1980 Vietnam had utilized satellite data provided by the United Nations Development Programme, but not until 1998 was the first national satellite project approved by the Vietnamese government. But the process of navigating the ins and outs of satellite registration with the International Telecommunication Union (ITU) took an entire decade to work itself out. In 2002, all relevant industries and ministries in Vietnam were asked to prepare for a national space program. A call for bids to build the satellite was issued and three bids were considered, from Lockheed Martin (USA), Alcatel Alenia Space (France), and Sumitomo Corporation (Japan).[111] Lockheed Martin was contracted in 2006, and on 18 April 2008 the US$200-million *Vinasat-1* lifted off from the Guiana Space Center, making Vietnam the sixth southeastern Asian country to own a satellite. The Canadian satellite firm Telesat was contracted to provide engineering support.

As a communications platform, *Vinasat-1* was intended to give Vietnam telecommunications autonomy and free up an estimated US$10–15 million in annual leasing fees that was previously paid for access to other countries' telecom satellites. Besides providing universal television and internet coverage to Vietnam, *Vinasat-1* was expected to become a vital asset in bolstering the production of Vietnam's important fishing industry by providing GPS information as well as weather and rescue data. Two satellite control stations have been constructed: one in the northern Ha Tay province and one in the southern Binh Duong province. The follow-up US$350-million *Vinasat-2* communications platform (also built by Lockheed-Martin) is scheduled to be put in orbit in the second half of 2012 from the Guiana Space Center. It is noteworthy that the Vietnamese government has been forthright in its view that these satellites are an important symbol of Vietnam's "orbital sovereignty" and placeholders for the country's increasing important economic position in the international system, thus emphasizing the continued role of space as an international prestige creator.[112]

In June 2006, the Vietnamese government approved a national strategy for space technology research and applications for the 2006–20 period, which called for: (1) the establishment of a legal framework and a policy for cooperation with international partners; (2) the construction of a space industry infrastructure; (3) an educational plan to enable indigenous space technology manufacturing; (4) the manufacture of satellites; and (5) the application of space technology for the socioeconomic development of the country.[113] To achieve these goals, Vietnam's Space Technology Institute was founded in 2007 as one of 19 subdivisions of Vietnam's expanding Academy of Science and Technology. Its role will be to design and assemble small satellites, thus adding another degree of autonomy to the country space designs as part of the government National Strategy. In 2010, construction began on the US$600-million Hoa Lac Space Center, 30 kilometers west of Hanoi, a complex that will house a satellite assembly facility, an integration and test facility, a satellite signal transmission station, a research center, a space museum, and an observatory.[114] The facility is planned for completion by 2018.

Indonesia: more than 17,000 reasons to go into space

In contrast to the relatively recent appearance of many EMSAs, Indonesia's official space policy is of long standing. Founded in 1964 by the former president Suharto, the National Institute of Aeronautics and Space (Lembaga Penerbangan dan Antariksa Nasional—LAPAN) is the oldest national space agency in Asia and among the world's oldest. A member of the International Astronautical Federation, LAPAN has had the mission of promoting both civilian and military aerospace activities. But with a paltry operational budget of US$20 million, it is easy to understand why LAPAN has maintained historically close ties to the Indonesian military (Tentara Nasional Indonesia—TNI). Thus, as in China, the genesis and *raison d'être* of LAPAN is found, in part, in the development of rocketry for military applications. Regardless of its military liaisons, LAPAN has also played an important role in socioeconomic development through the LAPAN Earth observation program, which uses satellite data to mitigate natural disasters (Indonesia is located in one of the world's most active seismic zones) as well as for environmental monitoring, particularly against forest fires, illegal logging, and slash-and-burn agriculture.

Indonesia's first satellite, the *Palapa A1* (Indonesian for "Fruits of Labor"), was launched from Cape Canaveral in 1976. Built by Boeing, *Palapa A1* was Indonesia's first communications platform and the first satellite operated domestically by a developing country. The follow-up, *Palapa A2*, was launched the next year to begin the creation of Indonesia's indigenous satellite telecommunications system, which was designed to provide telephone and television service for the inhabitants of the over 17,000 islands in the country. From 1976 to 2009, Indonesia launched 13 different telecommunications satellites in the *Palapa* series, creating the best communications systems in the region.

Indonesia's first remote-sensing satellite, *LAPAN-TUBSat*, was launched from India's Sriharikota space center on 1 January 2007. As a joint project between LAPAN and the Technical University of Berlin, *TUBSat* represented the expansion of Indonesia's capabilities in weather forecasting and national resource location; moreover, its five-meter resolution camera provided the TNI the ability to monitor gas fields near the Ambalat sea block in the Celebes Sea, which neighboring Malaysia also claims.[115] Indonesia's newest satellite entries were launched in 2011 from India's Sriharikota launch site. Built as a joint venture between LAPAN and the Indonesian Amateur Radio Organization, the twin remote-sensing satellites *Lapan-A2* and *Lapan-Orari* were put in an equatorial orbit instead of a polar one by an Indian PSLV-CA launcher to provide longer coverage time for the country's 17,508 islands. Among the planned missions of the satellites will be the use of their three multiband spectral imaging cameras to identify forest fires, a constant threat in the country due to slash-and-burn agriculture, as well as to provide disaster support.[116]

While making a name for itself as a prolific user of satellite technology for socioeconomic purposes, Indonesia held high ambitions to complement its growing armada of satellites with an equivalent indigenous launch capability. As early as 1984, Indonesia began launching sounding rockets from the Pameungpeuk launch site, located about 170 kilometers southeast of Jakarta. The solid-fuel rockets reached between 150 to 300 kilometers in altitude. The country's first full-fledged experimental launch vehicle, the five-meter-long RX-250, rose from Pameungpeuk in 1987 to an apogee of 70 kilometers.

With the experimental stage over, LAPAN has taken the important first step toward Indonesia's acquisition of independent launch capability. In 2008, LAPAN undertook a new missile development project along with the TNI. The ultra-light RPS (Roket Pengorbit Satelit—Satellite Orbiting Rocket) project is expected to produce a three-stage launcher capable of lifting a microsatellite into orbit. In 2009 the 6.2-meter RX-420 experimental rocket was successfully test-launched from Pameungpeuk and reached an altitude of 50 kilometers. The RX-420 tested the technologies and techniques that will allow Indonesia to become launch-autonomous by 2015. The final version, the RX-520, is scheduled for testing in 2012.[117] In the meantime, Indonesia has been actively strengthening space relationships with more advanced space actors to augment its experience and technology, particularly with the United States and China. In October 2010, China announced that it will assist Indonesia in the latter's RPS program and will make available opportunities for Indonesian astronauts to become involved in Chinese space flight programs.

Bangladesh: not too poor for space

Since separating from Pakistan following the 1971 Liberation War, Bangladesh has eagerly sought ways to mitigate the challenges presented by its geography and socioeconomic situation. The vast majority of the country's 110 million inhabitants live in an area prone to flooding, cyclones, tornadoes, and storm surges. Since the 1930s, aerial photography had been used for mapping and forestry purposes, but space age technologies have greatly enhanced these capabilities. The Space Research and Remote Sensing Organization (SPARRSO) was established in 1980 under the Bangladeshi Ministry of Defense, though in this instance, not too much should be read into the institutional placement.

As is very common in poorer developing countries, the military has traditionally been more involved in national development projects because it typically is one of the only national entities that can provide both the necessary organizational skills and the trained and educated personnel. The SPARRSO has been charged with developing and implementing the peaceful applications of space science, but because of its utter lack of infrastructure, Bangladesh has sought to form partnerships with more developed space actors. In September 2010, the Bangladeshi government announced its intention to invest US$200–300 million in a telecommunications satellite in order to bring Bangladesh's broadcasting facilities up to international standards.[118]

Pakistan: between India and a hard place

Since its divorce from India in 1947, Pakistan has found itself in a self-imposed dilemma. As in Israel and North Korea, Pakistan's drive to build and improve its missile and subsequently space technology must be understood largely in the geopolitical context and, also like Israel and North Korea, Pakistan often shrouds the improvement in its ballistic missile and nuclear programs within the context of a public civilian space program. As has been discussed, many space-based and satellite systems are inherently dual-use technologies, and in Pakistan this lack of distinction is especially pronounced because of the country's precarious strategic position. For this reason, Pakistan's civilian space program has been intimately linked to its military strategic plans.

Having fought India in three major wars (1948, 1965, and 1971), one undeclared war, and numerous border skirmishes, Pakistan's MNS triad may be correctly understood as being almost totally oriented toward maintaining strategic parity with India. The Space Sciences Research Wing of the Pakistan Atomic Energy Commission (PAEC) was established in 1961, and on 7 June 1962, Pakistan test-fired a NASA *Nike-Cajun* sounding rocket (renamed *Rehbar-I*) from the Sonmiani Rocket Range; it climbed to an altitude of 130 kilometers. Pakistan thus became the third Asian country to put a rocket in space. Two days later *Rehbar-II* repeated the feat. Soon thereafter, Pakistan put into operation its own sounding rocket program, and some 200 sounding rockets have been launched from the Sonmiani Beach flight range, approximately 145 kilometers northwest of Karachi.

The coordination of Pakistan's nascent space program was established in 1961 with the administrative founding of three agencies: the Space Research Council (SRC), an Executive Committee of the Space Research Council (ECSRC), and the Pakistan Space and Upper Atmosphere Research Commission (SUPARCO). The first administrator of SUPARCO and architect of Pakistan's aerospace industry and space program was Władysław Józef Marian Turowicz, a Polish-born aeronautical engineer who had fought for the British in Pakistan during World War II and settled there after the war. Headquartered in Karachi, SUPARCO is charged with the development of missile and satellite technology and implements space policy created by Pakistan's Space Research Council (SRC), and states that its primary mission is Earth imaging and upper atmospheric research.[119] It was divided into three areas: Space Technology, Space Research, and Space Electronics. From 1981 to 2000, the institutional structure of

SUPARCO was placed under the direct control of the Pakistani president. However, following the military coup and overthrow of the civilian government in October 1999, the Space Research Council and the Executive Committee of Space Research Council were dissolved and were replaced by the Development Control Committee of National Command Authority, a new military organization responsible for national strategic planning, including the country's nuclear forces.

As mentioned earlier, Pakistan's achievements in satellite and launcher technologies are inexorably tied to its nuclear program and the ongoing rivalry with India; these two security motivators have ultimately driven Pakistan's space program. All aspects of the MNS triad in Pakistan have had one goal: to maintain parity with India, if not surpass it. The genesis of this policy began after Pakistan's defeat in the 1971 war with India, when the decision was made to create a nuclear weapons program. The next year, the Minister for Fuel, Power and Natural Resources (and future president) Zulfiqar Ali Bhutto founded the Pakistani nuclear program. India's nuclear test in 1974 just made obvious what Islamabad had until then tried to keep secret: nuclear weapons were to be pursued if Pakistan hoped to remain its sovereignty vis-à-vis India.

Though Pakistan had run a civilian nuclear program since 1956 when the PAEC was established to take advantage of the US Atoms for Peace Program, and a small research reactor opened near Islamabad in 1960, nuclear weapons became the sole focus of the program after 1974. Under the Directorate of Technical Development, an enrichment plant was opened in Chaklala and a reprocessing center was designed by the Belgian firm Belgonucléaire.[120] Pakistan was aggressively intent on completing the MNS triad. By 1979 the Kahuta uranium enrichment plant was producing uranium under the direction of Abdul Qadeer (A.Q.) Khan, the father of Pakistan's nuclear program, who hastened its birth using stolen designs from the Dutch research laboratory he had previously worked for.

At the same time during the 1970s, Pakistan began building its own rocket engines and setting the stage for rapid progress toward its future space activities. SUPARCO began utilizing remote-sensing data from US Landsat and NOAA satellites in 1973 and built a formal remote sensing analysis center called RESACENT in 1978 to disseminate data to policy-makers for various projects in water management, forestry, and land use.[121] Through the 1980s, SUPARCO also honed its rocketry skills through the manufacture of battlefield rockets, such as the short-range *Hatf* series, which began production in 1989. To complement its maturing missile systems, after a period of ambiguity during the 1980s (mirroring or even imitating Israel's policy), Pakistan emerged as both a nuclear and missile power. During the 1980s a space launcher factory was built, as were rocket-testing facilities, radar tracking, and telemetry-receiving facilities. In yet another echo of Kennedy's "going to the moon" speech, the director of SUPARCO, Salim Mehmud, declared in 1981 that Pakistan would launch an indigenous satellite launch vehicle within a decade.[122] To achieve this goal, Pakistan became a steady customer of Chinese technology, to such extent that Pakistan has been described by the Chinese military as "China's Israel" because of the dependency relationship.[123]

Pakistan did acquire its first indigenous satellite within the decade, the *Badr-1* ("New Moon-1" in Urdu), but not on its own. The satellite was originally slated for launch aboard the US space shuttle *Challenger* in 1986, bit the shuttle's destruction and the subsequent suspension of shuttle operations forced Pakistan to look for alternatives. The satellite was finally launched into LEO in July 1990 aboard a Long March LM-2E from China's Xichang Satellite Launch Center. Like most first satellites, *Badr-1* was largely experimental in nature and provided Pakistani engineers valuable experience with satellite-related technologies such as telemetry as well as voice and data communications. The follow-up *Badr-2* was an Earth observation platform co-designed by SUPARCO and the British company Space Innovations Limited. *Badr-2* was originally planned to become the first Pakistani launched indigenously from the Tilla Satellite Launch Center in Punjab Province, a launch center jointly run by SUPARCO and the Pakistan Atomic Energy Commission (PAEC). But because of political instability caused by the consolidation of Musharraf's rule, the 70-kilogram satellite was instead launched abroad a Ukrainian Zenit-2 rocket and placed into sun-synchronous orbit from Russia's Baikonur Cosmodrome on 10 December 2001.

Completing the second leg of the MNS triad with its yet-untested nuclear weapons capability by 1987, Pakistan had only US-purchased F-16 fighters at its disposal as a means of nuclear delivery. In the early 1990s, short-range M-11 Dong Feng missiles (renamed *Shaheen* in Pakistan) and a new missile assembly plant were purchased from China. But in June 1991 the United States imposed sanctions on both China and Pakistan for the proliferation of M-11 missile technology and components. Pakistan was forced to find more surreptitious sources to improve its missile technology. A.Q. Khan again became an active interlocutor for the Pakistani nuclear and space programs. The intensification of the US-led nonproliferation regime ultimately forced Pakistan's space program into a completely new orbit.[124]

Using A.Q. Khan's connections in North Korea (to whom he had been secretly selling nuclear designs), Pakistani Prime Minister Benazir Bhutto made a cash deal with North Korea for its longer-range *Nodong* missile, which was renamed Ghauri.[125] In addition, North Korea sold Pakistan 15 metric tons of ammonium perchlorate (an oxidizer in solid rocket fuel), though it was confiscated in Hong Kong in April 1996.[126] In April 1998, Pakistan tested the *Ghauri*, whose 2,300-kilometer range meant that almost every city in India was now within striking distance. This development forced the Indian government to react for domestic political reasons. India rushed ahead with five nuclear weapons tests in early May 1998

(Operation Shakti), which brought in response six Pakistani nuclear tests two weeks later (Operation Chagai-I). The subcontinent's nuclear weapons status was now crystal-clear. This nuclear tit-for-tat game pushed ahead plans for Pakistan's development of a space-worthy launcher.

The latest chapter in Pakistan's trek into space coincides with political upheaval. After General Pervez Musharraf's non-violent seizure of power in 1999, the Pakistani military assumed control over all aspects of national security and national development. Musharraf created a National Security Council (NSC), which gave the military the last word in Pakistani policy-making. The NSC was subdivided into three sections: the National Command Authority (NCA), the Strategic Plans Division, and the respective strategic force commands of the three military services (Army, Air Force, Navy).[127] By charter, the NCA is chaired by the Pakistani president, who had also become the country's military leader.

As a result of this organizational overhaul, the scope and parameters of the space program were expanded. The Space Applications Research Center was commissioned in 1999 at Lahore, and the Aerospace Institute, under SUPARCO, began programs to train space application experts. Existing ground control stations that receive data from US Landsat, NOAA, and SPOT satellites were renovated and the national Geographic Information Systems committee was standardized to international norms. In November 2000, Pakistan's nuclear and missile programs were put under the control of the NCA, which included SUPARCO because of the new government and the space agency's historically close ties to the military. After meetings with A.Q. Khan, Musharraf also took SUPARCO under his wing, saying "SUPARCO has suffered severe economic and global sanctions but in future Pakistan will send its satellites from its soil."[128] The once-moribund space program was breathing again and had received a mandate to build a satellite launcher. In 2005 Musharraf outlined a long-term vision for the Pakistani space program, saying that given Pakistan's existing nuclear and missile programs, the country's graduation into space technology must be the logical next step and that SUPARCO must deliver.[129]

The product of SUPARCO's new mission was unveiled on Pakistan Day in March 2000. The *Shaheen II* is Pakistan's response to India's *Agni II*. The *Shaheen II* could essentially reach all of India, forming the backbone of a Pakistani nuclear deterrent. It was successfully test-fired in March 2005 and its dual-use design allows it to not only carry conventional and nuclear warheads, but also place a satellite into orbit. Concurrently, Pakistan's first communications satellite, the Hughes-manufactured *Paksat-1*, was purchased for US$4.5 million. Originally Indonesia's satellite (known as *Palapa*), it had suffered an electrical failure. Pakistan quickly acquired the faulty satellite to hold the orbital slot (38E) that had been allotted to it (under the UTI's three-year "use it or lose it" provision). Once repaired, *Paksat-1* was maneuvered into position over Pakistan in 2002, and provides communication services to over 75 countries in the Middle East, Europe, Africa, and south-central Asia. The civilian applications of *Paksat-1* include a reduction in the previously prohibitive cost of leasing bandwidth, which had prevented Pakistan from using satellite technology for education. Four education channels were part of an inaugural Distance Education Program carried via *Paksat-1* to the more remote parts of the country.[130]

In 2009 China and Pakistan signed an agreement for the promotion of satellite technology worth US$222 million. The Chinese Great Wall Industry Corporation-manufactured *Paksat-1R* communications satellite was launched from China's Xichang Satellite Launch Center on 11 August 2011. A subsequent satellite will be manufactured at the SUPARCO center in Lahore, making it the country's first indigenously developed remote-sensing satellite for agriculture, crop monitoring, yield estimation, food and water security, improvement of water courses, monitoring of the environment, disaster monitoring and mitigation, and land cover use.[131] In 2011, Pakistani Prime Minister Syed Yusuf Raza Gilani announced the Space Vision 2040 plan, which established the utilization of space technology as an instrument of socioeconomic development and national security. Also in the works is Pakistan's indigenous Satellite Launch Vehicle (SLV), which is slated for debut in 2012.

Malaysia: space is the place to be

Malaysian formal space activities are recent, but reflect a highly developmental approach to the use of space activity to serve national interests. Originating as the "Planetarium Division" of the prime minister's office in 1992, Malaysia's modern space agency, the National Space Agency of Malaysia (ANGKASA), was founded only in 2002. It has been charged with a broad-reaching and ambitious set of goals, which include fostering the development of indigenous space technologies, assisting the government in formulating space policy, and also providing inspiration for younger Malays to become more scientifically savvy.

The primary focus in the application of Malaysian space technologies has been on building a satellite fleet to provide for its communication and scientific needs in the service of national development. As early as the 1970s, Malaysia was on the forefront in Asia in utilizing remote-sensing data (principally from the US and France) to monitor its valuable and endangered tropical forests.[132] The Malaysia Centre for Remote Sensing was founded in 1988 as an agency under the Ministry of Science, Technology and Innovation of Malaysia to coordinate data from remote sensing for applications in agriculture, forestry, hydrology, topography, and socio-economic areas. Forest fires are one area in particular where remote sensing has been seen as invaluable, given the fires' frequency and propensity to cause significant health and economic problems.

To properly plan the country's investment in space program, the white paper *National Blueprint for the Malaysian Aerospace Industry* was issued by the Malaysian government in 1996. It lays out a design to foment an aerospace industry that will transform the country into a space actor by 2015.[133] As a tangible first step, the Malaysia International Aerospace Centre (MIAC) was created at the Subang airport to promote aerospace training and technology development. Another demonstrable step toward space was the establishment of the country's MEASAT constellation of geosynchronous satellites, built by the Boeing Satellite Systems and launched in 1996 by an ESA *Ariane* from French Guiana. Malaysian operators control the satellites from a facility on Langkawi Island.

Malaysia's first satellite, the *MEASAT-1*, is a communications platform designed and constructed by Hughes Space and Communications International and launched in January 1996 from French Guiana, as was its sister satellite, *MEASAT-2*, in November 1996. The third in the series, *MEASAT-3*, was put in orbit from Baikonur in December 2006, creating one of the more extensive telecom networks in Asia. In 2000, *TiungSAT* was launched from Baikonur Cosmodrome in Kazakhstan. It was Malaysia's first microsatellite and was jointly developed with Surrey to conduct experiments in remote imaging and science.

Perhaps the most notable and interesting aspects of Malaysia's space program is its early emphasis on manned flight. While the country is not yet launch-capable, the Malaysian government has approved the Angkasawan ("astronaut" in the Malay language) program, which was established as an in-kind agreement between Russia and Malaysia: Russia would pay for the training of two astronauts (a primary and a back-up) and then fly one of them to the International Space Station (ISS) for a 10-day stay, in exchange for Malaysia's US$1-billion purchase of 18 Russian Sukhoi SU-30 fighter jets.[134] Though not the first Muslim in space (eight had preceded him; the first was Saudi prince Abdul Aziz Al-Saud aboard space shuttle *Discovery* in 1985), Sheikh Muszaphar Shukor's October 2007 flight highlighted the challenges that Muslims confront in practicing their religion in space, such as finding Mecca to face for prayers. A resulting publication, "Guidelines for Performing Islamic Rites (Ibadah) at the International Space Station," was published by the Malaysian government, suggesting that each Muslim astronaut figure out the best way according to each one's ability.[135]

Taiwan: development and defense combined

Like Israel, Taiwan does not qualify under traditional economic definitions as a developing country, but it is definitely a rising space actor whose space policy has been chiefly responsive to the challenges and needs of development and to the rising space programs of other Asian space programs. Taiwan has been aggressively pursuing a small but ambitious space program based solely on a developmental goal but with a strong independent streak, which is perhaps not too surprising given its tenuous relationship with its largest neighbor, China.

Taiwan's National Space Organization (NSPO) was originally founded in October 1991 as the National Space Program Office. The NSPO is currently in the first phase of a 15-year space program called "Space Technology Long-Term Developmental Program" (STLTDP), which seeks to build space infrastructure and industries on the island state. As part of the first phase, launcher development efforts have proceeded apace. Taiwan's first sub-orbital sounding rocket, the eight meter SR-1, was first launched in December 1998 from Jiu Peng Air Base in southern Taiwan. Six more sounding rockets followed with a specific goal—to serve as the foundation for an eventual indigenous satellite launch vehicle.[136] In the spirit of this goal, all components of the SR series have been indigenously produced at Taiwanese universities. The second phase of the STLTDP began in 2004 and includes the ambition of a home-made satellite launcher.

Taiwan's first satellite, *Rocsat-1* ("Republic of China Satellite-1"), was built by TRW, Inc. of Cleveland, Ohio, with the cooperation of Taiwanese scientists for the purpose of technology transfer. The multi-purpose satellite, later renamed *Formosa-1* ("Beautiful," from the old Portuguese name for the island), was launched from Cape Canaveral on 27 January 1999. Two more in the series, *Formosa-2* and *Formosa-3*, followed in 2004 and 2006. The next in the series promises not only to expand Taiwan's satellite communications and remote-sensing capabilities, but will perhaps open a new chapter in the history of space programs.

The NSPO has contracted Space Exploration Technologies (SpaceX) as a private launch provider for *Formosa-5*, which is scheduled to lift off from its site in the Kwajalein Atoll some 4,000 kilometers southwest of Hawai'i in 2013. Taiwan's other indigenous satellite effort was the ESEMS (Experimental Scientific-Education Microsatellite), which was produced jointly by two Taiwanese universities and Lomonosov Moscow State University and launched atop a *Soyuz-2–1B* from Baikonur in September 2009.[137] Taiwan was also one of 16 countries to contribute to the AMS-02 (Alpha Magnetic Spectrometer-02) experiment, which was lifted into space in May 2011 by the space shuttle *Endeavor* and installed on the International Space Station. The AMS-02 will study cosmic rays and dark matter in order to better understand the origin and evolution of the universe.

Other Asian space actors

A number of other Asian states have made strides toward utilizing space, primarily for telecommunications. Under the direction of the Philippine Aeronautics and Space Agency, the Philippines purchased a communications satellite from Space Systems/Loral in California. The *Agila-2* ("Eagle-2" in Tagalog) was launched aboard a *Long March* 3B from China's

Xichang Satellite Launch Center in 1997 and provides telephone, digital broadcast TV, and data services to the Philippines and Southeast Asia for the Mabuhay Philippines Satellite Corporation.

Thailand's entrance into the space club began with a more detailed plan. After many failed attempts over the years, in early 2004 the Thai Ministry of Information and Communication Technology (MICT) commissioned Chulalongkorn University in Bangkok to formulate a national space master plan. The proposed space budget was to begin at 0.01 percent of national GDP in 2006 and to progress to 0.05 in 2015.[138] With the launch of its remote-sensing *Theos* satellite aboard a Dnepr launch vehicle from the Dombarovskiy launch facility in southern Russia in 2008, Thailand joined its fellow southeastern Asian states in using space for development. The *Theos* was built by EADS-Astrium under a contract from the Thai Ministry of Science and Technology. Thailand's main space organization, the Geo-Informatics and Space Technology Development Agency (GISTDA), was founded in 2002 and is charged by the Thai government with the mission of the development and application of space technologies to "develop cooperative national and international networks to support sustainable development and improvement of natural resources and environment as well as quality of life."[139] The GISTDA coordinates all Thai space activities involving remote-sensing satellites.

Regional cooperation among EMSAs

The past two decades have seen a trend emerge among the EMSAs that, once again, mirrors the path blazed by the developed space actors. Regional cooperation, either ad hoc or through formal regional organizations, has become a mechanism for states to engage in a variety of activities that support the growth of the culture, technology, and infrastructure that promote space activities, even if the enormous capital investment requirements are lacking.

Regional cooperation in space activities has been attempted in Latin America. In 1991, a conference was convened in Costa Rica by Costa Rican-American astronaut Dr. Franklin Chang Díaz (who flew seven times on US space shuttles) to discuss the possibility of creating a Latin American space agency to coordinate the region's space-related activities. Conceived to be a counterpart of the European Space Agency, the group did not become a reality because of the objections of Brazil, which already possessed far superior technology and facilities to those of the other Latin American countries and saw no reason to diffuse its power through such an organization.

But inter-state cooperation has been accomplished on certain goal-specific projects. In 1993, six Latin American states—Argentina, Brazil, Chile, Costa Rica, Mexico, and Uruguay—collaborated in the Chagas Project, which created an experiment, taken aboard the space shuttle, to study the causes of the tropical parasitic disease Chagas. A subsequent regional preparatory meeting for Latin America took place in Concepción, Chile in 1998. One of the chief recommendations that came out of discussions was the creation and strengthening of governmental institutional mechanisms which would allow suitable development of space activities, with an eye toward providing solutions to the region's socioeconomic malaise. Currently, Brazil and Mexico cooperate in the Regional Center for Teaching and in Science and Space Technology for Latin America and the Caribbean (CRECTEAL), to share educational and technical knowledge throughout the region.

There have been further attempts to broaden, deepen, and institutionalize space programs in Latin America. Under the auspices of the United Nations Committee on the Peaceful Uses of Outer Space, the Regional Centre for Space Science and Technology Education for Latin America and the Caribbean (CRECTEALC) was established in São Paulo, Brazil in 2000 and an instructional center set up in Mexico in 2002. This center is specifically created to foster education and research on space-oriented topics and technologies in the region. In correlation to their relative expertise, representatives from Brazil and Mexico were chosen to head the organization (UNCOPOUS, 2006). However, by 2011 the mood for space cooperation had seemingly changed among the region's most advanced space actors, as Brazil and Argentina defense ministers have openly discussed the advantage of creating a South American space agency for cost-sharing as well as for mutual defense strategies, which will increasingly depend on space-based assets.[140]

Likewise, Asia has undertaken regional integration in space activities, though at a much higher level of institutionalization. Headquartered in Beijing, the Asia-Pacific Regional Space Agency Forum (APRSAF), as its name suggests, is an inter-governmental discussion that has been occurring annually since 1993. Spearheaded by the Japanese space agency (JAXA) and by China, the APRSAF brings together more developed (including Australia) and lesser developed Asian countries to work toward two goals: improving each participating country's space program and providing opportunities for cooperative space ventures, such as a disaster management system called Sentinel Asia for the Asia-Pacific region utilizing Asian-produced Earth observation satellite data, a project that commenced in October 2006.[141] Aside from these stated public goals, however, the underlying rationale for the APRSAF has been to ensure Asia's place as a leader in the future of space activities, thus bespeaking the role of space policy as an arm of national power, though in a liberal, cooperative framework.[142]

A similar forum and trajectory have been initiated in Africa. As early as 1991, Article 63 of the Abuja Treaty committed the African Union to the establishment of a pan-African satellite communications system. Though unrealized, it set the stage for an official intergovernmental forum on space cooperation. Co-sponsored by the Algerian Space Agency and United Nations Office for Outer Space Affairs (UNOOSA), the African Leadership Conference on Space Science and Technology for Sustainable Development has sought to accomplish much the same goals as APRSAF. Beginning in 2006, the main African EMSAs of Algeria, Kenya, Nigeria, and South Africa have used this forum to come to an agreement on the uses of space technologies in the areas of disaster management, resource identification, land use, and public health. The ultimate goal of an all-Africa satellite constellation to attain these goals has been proposed.[143]

The newest cooperative space organizations to emerge in the developing world are two found in the Muslim world. In 1986, the Inter-Islamic Network on Space Sciences and Technology (ISNET) was founded by nine Muslim countries: Bangladesh, Iraq, Indonesia, Morocco, Niger, Pakistan, Saudi Arabia, Tunisia, and Turkey. Since its founding, five more countries—Syria, Iran, Azerbaijan, Sudan, and Senegal—have joined. With its headquarters located in Pakistan's SUPARCO, the ISNET's stated objectives include cooperation in the peaceful use of space, information exchange, the promotion of cooperative activities, and the training of personnel in space technology.

The second is the proposed Pan-Arab Space Agency (PASA). The United Arab Emirates (UAE), whose government has expressed a keen interest in space activities, has hosted annual conferences since 2007 to explore and explain the value of space programs to the national development regional states. A central component in the goals of the PASA is the planned Arab Space Research Agency, which would merge training, knowledge, and technology between the predominantly Muslim countries of the Middle East and North Africa. Preliminary plans have already been submitted to governments across the region. Telecommunications, space science, and international participation are reasons cited by UAE for space investment, though other concerns have value as well. One is the geopolitical logic of the military use of space technology for states hoping to counter the growing influence of Iran in the region. Another longer-term reason is the stark realization that the Middle East's petroleum-based economy will not last forever (non-viable before 2050, by some estimates), thus the imperative for diversification into other high technology industries.

Conclusion
Space policy in developing countries

The evolution of space policy in the twenty-first century continues along the well-worn path blazed in the twentieth century. This earlier period was replete with exciting, even heady developments in rocket, satellite, optical, and remote-sensing technologies. After a beginning with the first almost primitive, grainy images of Earth's upper atmosphere shortly after World War II, today's modern states are now utterly dependent in countless ways on the fruits of space-based assets. Space-based communications, weather forecasting, high-resolution imagery, and orbital science have all become interwoven into the sinews of modern societies. Thus, they have become the touchstones upon which any capable state's modernity is measured.

However, as this book has examined, the evolution of the space programs that have made these technologies possible displays a notable pattern. For developed and developing countries alike, the decision to invest in a national space policy has been predicated not only on the desire for obtaining these technological advances but, when the financial and geopolitical circumstances allow and/or dictate, to address the perceived national security needs of the state. In fact, since the dawn of the space age, the one constant that can be discerned is the very strong tendency among such space-capable states to attempt to utilize space assets to their fullest measure in the service of national security as well as socioeconomic development. This approach fits well with our understanding of how other technological innovations have been exploited to the service of the state throughout the history.

While it is true that not every space-capable state has pursued a purely security-oriented space option, the intervening variable has frequently been a state's perception of its position vis-à-vis other states in the international system of the time. For example, while Canada, West Germany, and Japan were all technologically capable of completing the MNS triad during the Cold War, their relative positions as junior partners in a US-led system meant that pursuing more offensive options (especially nuclear) was not necessary, nor were these options feasible given the superpower dynamics of the time. But among those space-capable states that perceived their relative power position as less secure or desirable within the system, the completion (or attempted completion) of the MNS triad has normally been the result. Among developed states, post-World War II France best represents the archetype of this trend, with De Gaulle's ardent pro-nationalist policies leading France's expansion of the MNS triad as an integral part of an independent French space policy.

As the countries of the developing world emerged from the shadows of Cold War superpower hegemony, many embraced and adopted space policies of their own. All too often, the path that they chose to follow (or felt they were forced to follow) was analogous to the one blazed by the first space actors. Among the largest and wealthier emerging space actors, this pattern has manifested itself very consistently. When it comes to demonstrating national capacity, fomenting international prestige, and concretizing nationalist sentiments, space programs have become a time-honored choice among those states capable of undertaking the financial, technological, and political challenges. As this work has demonstrated, a plethora of developing countries as geographically and economically diverse as China, Argentina, Brazil, Iran, and South Africa all made the conscious policy decision to develop, or at least to attempt to develop, the MNS triad. Their reasons were as varied as their circumstances, but when space became a policy imperative, strategically oriented projects commanded the lion's share of allocated space budgets in these most capable EMSAs.

As might be expected, given the cost-intensive nature of space programs, the most significant constraint on the space programs of the EMSAs has been the money available to invest. But investment in space is growing apace as national governments continue to spend greater amounts on the next generation of space programs. This investment is expected to reach over US$70 billion and to grow by almost five percent annually.[1] Twice as many countries now dedicate state resources to space programs as did a generation ago, and the EMSAs are driving much of the growth.

For these reasons, the first tier EMSAs have naturally come from among the most economically advanced countries of the developing world. But concurrent with the economic means to pursue space has been the geopolitical necessity to pursue space-based assets for national security and prestige as well as the associated socioeconomic benefits. If a state wishes to be considered an emerging power, it must have the tools that the established powers already possess; this includes space programs. These government space policies have been driven by a state's long-term strategic national objectives. Those developing countries that have pursued missile and, occasionally, nuclear programs have invariably dabbled, if not invested heavily, in space-based applications. Thus, in these most advanced EMSAs it is the defense sector that has propelled the development of space technology in the past, and all indications are that either directly or indirectly, this association continues today, in many cases challenging the status quo established by the original space actors, the United States, Russia, and

Europe. As Deng Xiaoping understood for China, and as is true for all developing countries, science and technology are the chief productive force for growth and development on both the domestic and international fronts. In this pursuit, space programs are the ultimate achievement of that ideal.

Of equal importance has been the political will and foresight to employ space-based assets to benefit the state in both tangible and intangible ways. In this area, as has been described, not only have the most economically advanced countries of Brazil, China, and India developed space programs, but a host of medium-level powers have done so as well, though with proportionally scaled-down ambitions. Having a space program, like many technologically based necessities of the past, has become a matter of prestige and practical benefit that no capable state can or will do without. Since *Sputnik*, space activities have been a symbol of national pride as well as the most visible method for demonstrating technical expertise to the rest of the world. The wide swath of countries that have engaged in limited but determined space programs to address the needs of security and the attainment of prestige also look to space for benefits to improve their citizens' health, resource management, education, technological foundation, and communications. Many developing countries of the third tier have begun this same process through cooperative space programs with more developed space actors, both to improve their lot in the aforementioned areas and to acquire technical training and technology infrastructure development, much as NASA suggested over 40 years ago.[2] For the EMSAs of both the second and third tiers, the application of space-based assets still addresses the issue of national security, just from a less narrow scope than that employed during the earlier realist approach to the space race.

But in the end, the evolution of space policy in the developing world can be seen as merely another chapter in the ongoing story and evolution of the Westphalian system. Despite the tangible scientific, socioeconomic, and even military benefits, the important rationale of prestige, that oldest and most persistent of political virtues, must always be considered. Those countries that are capable of investing in space assets will continue to spend precious resources in order to produce the appearance of being more advanced and more like the developed space actors than their less space-fortunate neighbors or competitors. In the world of international diplomacy, appearances can make all the difference.

Thus, since the days the first rockets left the bonds of Earth, the possibilities of space have occupied a crucial place in the national policies of all capable nation-states. As the final frontier of the international system as we currently know it, space policy will continue to play an integral role in changing and perhaps equalizing the national security and development goals for the countries of the developing world in the twenty-first century.

Notes

Introduction: Space power as national power

1 Konstantin E. Tsiokovsky, *Beyond The Planet Earth* (transl. Kenneth Byers), Oxford, UK: Pergamon Press, 1960, 3.
2 *The Space Report: The Authoritative Guide to Global Space Activity*, Colorado Springs, CO: The Space Foundation, 2011.
3 "2010 State of the Satellite Industry Report Shows Continued Growth," Satellite Industry Association, 11 June 2008, available www.sia.org/news_events/pressreleases/Press_Release_StateotheSatelliteReport2010FINAL.pdf (accessed 18 June 2009).
4 "China Working on Nuclear-powered Lunar Rover," CBC News, 2 April 2007, available www.cbc.ca/news/story/2007/04/02/tech-rover.html (accessed 7 August 2007).
5 SATCAT Boxscore, Center for Space Standards, available www.celestrak.com/satcat/boxscore.asp (accessed 5 May 2008).
6 "2006 Commercial Space Transportation Forecasts," Office of Commercial Space Transportation of the Federal Aviation Administration, May 2006, 10, available www.faa.gov/about/office_org/headquarters_offices/ast/media/2006_Forecast_GSO_NGSO_May_24.pdf (accessed 14 August 2007).
7 European Policy Institute, *Yearbook on Space Policy, 2006/2007*, 2008, 42.
8 "Alternatives for Future US Space-Launch Capabilities," *Congressional Budget Office*, October 2006, 20, available www.cbo.gov/ftpdocs/76xx/doc7635/10–09-SpaceLaunch.pdf (accessed 20 June 2007).
9 US Congress. House. *Conference Report on H.R. 5631*, Department of Defense Appropriations Act, 2007. 109th Congress, 2nd Session, 2006.
10 Euroconsult, 18 December 2008, available www.euroconsult-ec.com (accessed 20 February 2009).
11 This consideration, for example, is what produced the US Abrams tank. The US military originally considered purchasing the (West) German Leopard tank, but decided against it so as not to be dependent on a foreign power, even an ally, for its main battle tank.
12 Geraci, "Army Transformation War Game," 2003, 5.
13 United States Secretary of Defense, "Memorandum to Secretaries of the Military Departments," 9 July 1999, 2, available www.au.af.mil/au/awc/awcgate/dod-spc/dodspcpolicy99.pdf (accessed 10 October 2007).
14 Budget of the Brazilian Ministry of Science and Technology, 2009, available www.camara.gov.br/internet/comissao/index/mista/orca/orcamento/OR2009/Proposta/projeto/volume4/tomo1/03_mct.pdf (accessed 21 June 2010); "India Boosts Space Budget," 15 March 2010, available www.space.com/8043-india-boosts-space-budget.html (accessed 21 June 2010).

15 Dean Cheng, "Pandas in Orbit: China's Space Challenge," Heritage Foundation, 8 October 2008, available www.heritage.org/events/2008/10/pandas-in-orbit-chinas-space-challenge (accessed 14 December 2009).
16 Vernon Van Dyke, *Pride and Power: The Rationale of the Space Program*, Urbana, IL: University of Illinois Press, 1964, 123.
17 Frank Morring, Jr., "Study Finds Launch Costs Dropping," *Aviation Week*, 5 May 2010.
18 World Bank, *Data & Statistics: Country Groups*, 2007.
19 *Human Development Report 2007/2008*. New York: United Nations Human Development Programme, 2007, 244–5.

Chapter 1   Space power and the modern state

1 Carl Sagan, *Pale Blue Dot*, 1994, 306.
2 Elhefnawy, "Rise and Fall of Great Space Powers," 2007.
3 Oberg, *Space Power Theory*, 1999, 10.
4 Sheehan, *International Politics of Space*, 2007, 2.
5 Nye, *Paradox of American Power*, 2002, 4–9.
6 C. Sagan, op. cit., 268.
7 *Exploring the Moon and Mars: Choices for the Nations*, Office of Technology Assessment, Washington, DC: Congress of the United States, 1991, 97–8.
8 Thompson, "Virtual Regime", 2002, 507–8.
9 Edward Carr, *The Twenty Years' Crisis, 1919–1939*, New York: Perennial, 2001; Zbigniew Brzezinski, *The Choice: Global Domination or Global Leadership*, New York: Basic Books, 2006.
10 *Draft Statement of US Policy on Space*, 3 April 1962, declassified 7 April 1982, available http://history.nasa.gov/SP-4407/vol. 5/ (accessed 11 November 2007).
11 Thomas S. Kuhn, *The Structure of Scientific Revolutions*, Chicago: University of Chicago Press, 1962, 5.
12 Gray and Sloan (eds.), *Geopolitics, Geography and Strategy*, 1999, 162.
13 Sheehan, op. cit., Chapter 1.
14 Joseph E. Justin, "Space: A Sanctuary, the High Ground, or a Military Mission?" RAND Occasional Paper, 1982, available www.rand.org/pubs/papers/P6758.html (accessed 21 September 2007).
15 Nicolas Peter, "The Changing Geopolitics of Space Activities," *Space Policy*, vol. 22, May 2006, 100.
16 Nye, op. cit., 140.
17 Dolman, *Astropolitik*, 2002 [esp. Chapter 4].
18 Hans Morgenthau, *Politics Among Nations: The Struggle for Power and Peace*, New York: Alfred A. Knopf Publishers, 1948.
19 Gearóid Ó Tuathail, "Understanding Critical Geopolitics: Geopolitics and Risk Society," in Gray and Sloan (eds.), *Geopolitics, Geography and Strategy*, 1999, 107.
20 Saul B. Cohen, *Geography and Politics in a Divided World*, London: Methuen, 1964, 24.
21 Gearóid Ó Tuathail, "Problematizing Geopolitics: Survey, Statesmanship and Strategy," *Transactions of the Institute of British Geographers*, vol. 19, no. 3, 1994, 260.
22 Hedley Bull, *The Control of the Arms Race: Disarmament and Arms Control in the Missile Age*, London: Institute for Strategic Studies, 1961, 175.
23 Dolman, op. cit., 8.
24 William H. McNeill, *The Pursuit of Power: Technology, Armed Force and Society Since A.D. 1000*, Chicago: University of Chicago Press, 1982, Chapter 1.
25 Norman Friedman, *Seapower as Strategy: Navies and National Interests*, Annapolis, MD: Naval Institute Press, 2001, 138.

26 Alfred Thayer Mahan, *The Influence of Sea Power upon History*, Mineola, NY: Dover Publications, reprinted 1987.

27 John M. Logsdon, "Human Space Flight and National Power," *High Frontier, The Journal for Space & Missile Professionals*, March 2007, 12.

28 Dolman, op. cit., 104.

29 Daniel Deudney, "Space after the Cold War," *Ad Astra*, February 1990, 3.

30 Solis Horwitz, "National Aeronautic and Space Act of 1958—S. 3609", NASA archives.

31 James Mazol, "Considering the FY 2010 National Security Space Budget," *Marshall Institute Policy Outlook*, July 2009, available www.marshall.org/pdf/materials/720.pdf (accessed 13 May 2010); Todd Harrison, "Analysis of the FY 2011 Defense Budget," Center for Strategic and Budgetary Assessments, 49, available www.csbaonline.org/wp-content/uploads/2010/06/2010.06.29-Analysis-of-the-FY2011-Defense-Budget.pdf (accessed 13 May 2010).

32 Russian Government Decree no. 635, 2006.

33 National Aeronautics and Space Act of 1958, Section 205, available http://history.nasa.gov/spaceact.html (accessed 20 May 2007).

34 Stephen D. Krasner (ed.), *International Regimes*, Ithaca, NY: Cornell University Press, 1983, 1.

35 An enlightening work on difficulties of the application of law in the open ocean, especially concerning modern-day piracy, is William Langewiesche, *The Outlaw Sea*, New York: North Point Press, 2004.

36 Arthur K. Kuhn, "Codification of International Law and the Fifth Assembly," *American Journal of International Law*, vol. 34, no. 1, January 1940, 104.

37 Good discussions of this problem are found in Reynolds and Merges, *Outer Space*, 1997, 11–15 and Dolman, op. cit., Chapter 5.

38 D.W. Moore, "Support for NASA Shuttle Flights Remains Firm," Gallup News Service, Poll Analyses, 17 February 2003, available www.gallup.com/poll/releases

39 Herbert E. Krugman, "Public Attitudes Toward the Apollo Space Program, 1965–1975," *Journal of Communication*, vol. 27, no. 4, 7 February 2006.

40 Susan Buck, *The Global Commons*, Mercer Island, WA: Island Books, 1998, 132.

41 *Declaration of the First Meeting of Equatorial Countries*, 3 December 1976, reproduced in *Journal of Space Law*, vol. 6, Spring/Fall 1978.

42 Joseph M. Grieco, "Anarchy and the Limits of Cooperation: A Realist Critique of the Newest Liberal Institutionalism," *International Organization*, vol. 42, no. 3, Summer 1988, 487–8.

43 *Exploring the Moon and Mars*, op. cit., 97–8.

Chapter 2 The evolution of national space policies

1 Scott D. Sagan, "Why Do States Build Nuclear Weapons?" 1996–97, 55.

2 Numerous writers have argued this point. See, for example, Dolman, *Astropolitik*, 2006, 86–7; Erickson, *Into the Unknown Together*, 2005, 105, 157.

3 Partington, *History of Greek Fire and Gunpowder*, 1999, 276.

4 A. Kosmodemyansky and X. Danko, *Konstantin Tsiolkovsky: His Life and Work*, Honolulu, HI: University Press of the Pacific, 2000, 36.

5 Guillermo O. Descalzo, "Pedro Paulet Mostajo: Un Visionario Peruno," *Air & Space Power Journal* (Spanish edn.), Third Trimester 2007, available www.air-power.maxwell.af.mil/apjinternational/apj-s/2007/3tri07/descalzo.html (accessed 30 March 2009).

6 Jacob Neufeld, *The Development of Ballistic Missiles in the United States Air Force, 1945–1960*, US Government Printing Office, 1990, 35.

7 Herbert A. Johnson, *Wingless Eagle: US Army Aviation through World War I*, Chapel Hill, NC: University of North Carolina Press, 2000, 142–3.

8 Robert H. Goddard, "A Method of Reaching Extreme Altitudes," *Smithsonian Miscellaneous Collections*, vol. 31, no. 2, 1919.

9 Michael J. Neufeld, *The Rocket and the Reich: Peenemünde and the Coming of the Ballistic Missile Era*, New York: The Free Press, 1994, 27–8.

10 "The Man Who Opened the Door to Space," *Popular Science*, May 1959, 128–9.

11 M.J. Neufeld, *Von Braun*, 2007, 75–8.

12 Quoted in "The V Weapons Campaign Against Britain, 1944–1945," Imperial War Museum, available http://london.iwm.org.uk (accessed 21 June 2008).

13 *Time*, 17 February 1958, 5.

14 M.J. Neufield, *Von Braun*, 170.

15 Dieter Hölsken, *V-Missiles of the Third Reich, the V-1 and V-2*, Sturbridge, MA: Monogram Aviation Publications, 1994, 239; Steven J. Zaloga, *V-2 Ballistic Missile, 1942–1945*, Oxford, UK: Osprey Publishing, 2003, 33.

16 Dieter K. Huzel, *Peenemünde to Canaveral*, Westport, CT: Greenwood Press Reprints, 1981, 33.

17 Hunt, *Secret Agenda*, 1991, 6, 21, 31, 176, 204, and 259.

18 Pavel V. Oleynikov, "German Scientists in the Soviet Atomic Project," *The Nonproliferation Review*, vol. 7, no. 2, Summer 2000, 4.

19 Hardesty and Eisman, *Epic Rivalry*, 2007, 23.

20 Lasby, *Project Paperclip*, 1975, 16.

21 Zahl, *Electrons Away*, 1968, 107; Lasby, op.cit., 251–52; Hunt, op. cit., 32–5.

22 Oleynikov, op. cit., 6–7.

23 Dickson, *Sputnik*, 2001, 64.

24 Linda Hunt, "US Coverup of Nazi Scientists," *Bulletin of the Atomic Scientists*, April 1985, 24.

25 Federal Bureau of Investigation, "Investigation of Arthur Louis Hugo Rudolph," 13 January 1949, available http://foia.fbi.gov/arudolph/arudolph1.pdf (accessed 9 August 2007).

26 Ronald C. Newton, *The "Nazi Menace" in Argentina, 1931–1947*, Palo Alto, CA: Stanford University Press, 1992, 376; Oliver Rathkolb, *Revisiting the National Socialist Legacy: Coming to Terms with Forced Labor, Expropriation, Compensation, and Restitution*, New Brunswick, NJ: Transaction Publishers, 2004, 219–23; Antonio C. Burgos, "Los 75 años de la Fábrica Militar de Aviones," *Aeroespacio* (Buenos Aires), September/October 2002, available www.aeroespacio.com.ar (accessed 12 September 2008).

27 David Sheinin and Beatriz Figallo, "Nuclear Politics in Cold War Argentina," *MACLAS Latin American Essays*, vol. 15, Mid-Atlantic Council on Latin American Studies, March 2001, 103.

28 K. Chatterjee, "Hindustan Fighter HF-24 Marut. Part I: Building India's Jet Fighter," Bharat Rakshak, available www.bharat-rakshak.com/IAF/History/Aircraft/Marut1.html (accessed 16 September 2008).

29 David Baker, *The Rocket: The History and Development of Rocket and Missile Technology*, London: New Cavendish Books, 1978, 103.

30 William P. Barry, "Sputnik and the Creation of the Soviet Space Industry," in Roger D. Launius, et al., *Reconsidering Sputnik: Forty Years since the Soviet Satellite*, London: Routledge, 2000, 96–7.

31 Peter A. Gorin, "Rising from the Cradle: Soviet Perceptions of Space Flight before Sputnik," in Launius *et al.* (eds.), *Reconsidering Sputnik*, 2000, 16–18.

32 J. Neufeld, op. cit., 24–34.

33 Philip Taubman, *Secret Empire: Eisenhower, the CIA, and the Hidden Story of America's Space Espionage*, New York: Simon & Schuster, 2003, 196.

34 J. Neufeld, op. cit., 37.

35 *Civilian Space Policy and Applications*, 1982, 306.

36 Donald Baucom, "Eisenhower and Ballistic Missile Defense: The Formative Years, 1944–1961," *Air Power History*, Winter 2004, 37.

37 Project RAND, *Preliminary Design of an Experimental World-Circling Spaceship*, Santa Monica, CA, 1946, 10, available www.rand.org/pubs/special_memoranda/SM11827.html (accessed 14 June 2008).

38 Steven Zaloga, "Most Secret Weapon: The Origins of Soviet Strategic Cruise Missiles, 1945–60," *The Journal of Slavic Military Studies*, vol. 6, no. 2, June 1993, 264.

39 Zbigniew Brzezinski, "The Cold War and Its Aftermath," *Foreign Affairs*, vol. 71, no. 4, Fall 1992, 5.

40 Neal, Smith, and McCormick, *Beyond Sputnik*, 2008, 21.

41 Dickson, op. cit., 68.

42 M. Brzezinski, *Red Moon Rising*, 2008, 89–90.

43 Norman M. Naimark, *The Russians in Germany: A History of the Soviet Zone of Occupation, 1945–1949*, Cambridge, MA: Belknap Press, 1995, 228; Ulrich Albrecht, *The Soviet Armaments Industry*, London: Harwood Academic Publishers, 1993, 88.

44 Firth and Noren, *Soviet Defense Spending*, 1998, 29; *Historical Tables: Budget of the United States Government*, Washington, DC: US Government Printing Office, 2004, 33.

45 "Contribution to the Soviet Space Program: Expenditure Implications of Soviet Space Programs," CIA Historical Review Program declassified document, April 1969, 2, available www.fas.org/irp/cia/product/sovmm69.pdf (accessed 14 June 2008).

46 Lawrence Freedman, *US Intelligence and the Soviet Strategic Threat*, London: Macmillan, 1977, 87.

47 Divine, *The Sputnik Challenge*, 29.

48 Linda T. Krug, "Presidents and Space Policy," in Sadeh (ed.), *Space Politics and Policy*, 2002, 64.

49 Asif A. Siddiqi, "Krolev, Sputnik, and the IGY," in Launius *et al.* (eds.), *Reconsidering Sputnik*, 2000, 46–7.

50 Daniel S. Greenberg, *The Politics of Pure Science*, Chicago: University of Chicago Press, 1999, 135.

51 Launius, "Eisenhower, Sputnik, and the Creation of NASA," 1996, 131.

52 Available www.thespacereview.com/archive/995b.pdf (accessed 5 June 2010).

53 Kenneth Osgood, "Before Sputnik: National Security and the Formation of US Outer Space Policy," in Launius *et al.* (eds.), *Reconsidering Sputnik*, 2000, 205.

54 Eisenhower, *Waging Peace*, 1965, 209.

55 McDougall, *The Heavens and the Earth*, 1985, 127.

56 McDougall, op. cit., 119; Hardesty and Eisman, op.cit., 62–3.

57 Divine, op. cit., 44.

58 Eugene M. Emme, "Presidents and Space," in Durant (ed.), *Between Sputnik and the Shuttle*, 1981,17.

59 Quoted in Z. Brzezinski (1992), 93.

60 Dwayne Day, "Cover Stories and Hidden Agendas: Early American Space and National Security Policy," in Launius *et al.* (eds.), *Reconsidering Sputnik*, 2000, 171–2.

61 Available http://history.nasa.gov/SP-4202/chapter6.html (accessed 26 July 2007).

62 George C. Sponsler, "The Military Role in Space," *Bulletin of the Atomic Scientists*, June 1964, 32.

63 S.N. Khrushchev, *Nikita Khrushchev*, 2001, 223.

64 William Taubman, *Khrushchev: The Man and His Era*, New York: W.W. Norton & Co., 2003, 358–60.

65 *Khrushchev Remembers: The Glasnost Tapes*, transl. and ed. Jerrold L. Schecter and Vyacheslav V. Luchkov, New York: Little Brown & Co, 1990, 188.
66 William Schneider, "Peace and Strength: American Public Opinion and National Security," in Gregory Flynn (ed.), *The Public and Atlantic Defense*, Lanham, MD: Rowman & Littlefield Publishers, 1985, 321–4.
67 J. Neufeld, op. cit., 18.
68 Ward, *Dr. Space*, 2005, 257.
69 Lyndon B. Johnson, speech to Democratic Caucus, Washington, DC, 7 January 1958.
70 Wheelon, "Lifting the Veil on Corona," 1995, 251; P. Taubman, op. cit., 217–18.
71 Handberg and Johnson-Freese, "Return of the American Military," 1997, 297.
72 Dwight D. Eisenhower, "Statement by the President Announcing the Successful Launching into Orbit of an Earth Satellite," 1 February 1958, *PPP, DDE*, 1958, Washington, DC: US Government Printing Office, 1959, 141.
73 L.T. Krug, *Presidential Perspectives on Space Exploration*, 1991, 26–7.
74 Pamela Ebert Flattau *et al.*, "National Defense Education Act of 1958," 2006, 24.
75 Roger L. Geiger, "What Happened After Sputnik? Shaping University Research in the United States," *Minerva*, Leiden: Springer Netherlands, vol. 35, no. 4, December 1997, 350–2.
76 *Discussion of the 357th National Security Council*, 6 March 1958, Dwight D. Eisenhower Presidential Library, available www.eisenhower.archives.gov (accessed 30 August 2007).
77 Erickson, op. cit., 8.
78 Quoted in Howard D. Belote, "The Weaponization of Space: It Doesn't Happen in a Vacuum," *Aerospace Power Journal*, Spring 2000, 46; Roger D. Launius, "An Unintended Consequence of the IGY: Eisenhower, Sputnik, the Founding of NASA," *Acta Astronautica*, vol. 67, nos. 1–2, July/August 2010, 255.
79 McDougall, op. cit., 194.
80 S.N. Khrushchev, op. cit., 87.
81 Ordway, Sharpe, and Wakeford, "Project Horizon," 1988, 1105–21.
82 Ward, op. cit., 171.
83 Wheelon, op. cit., 249–60.
84 Geoffrey Perret, *Eisenhower*, Avon, MA: Adams Media Corporation, 2000, 582.
85 P. Taubman, op. cit., 296.
86 William E. Burrows, *Deep Black: Space Espionage and National Security*, New York: Random House, 1986, 131.
87 I-Shih Chang, "Overview of World Space Launches," *Journal of Propulsion and Power*, vol. 16, no. 5, September/October 2000, 857.
88 Clayton K.S. Chun, "Expanding the High Frontier: Space Weapons in History," *Astropolitics: The International Journal of Space Politics & Policy*, vol. 2, no. 1, 2004, 66; Kenneth F. Johnson, "The Need for Speed: Hypersonic Aircraft and the Transformation of Long Range Airpower," Air University, School for Advanced Air and Space Studies, 2005, 18.
89 Curtis Peebles, "The Origins of the US Space Shuttle-1," *Spaceflight*, vol. 21, November 1979, 435.
90 Roy Houchin, *US Hypersonic Research and Development: The Rise and Fall of Dyna-Soar, 1944–1963*, New York: Routledge, 2006, 111.
91 Peter R. Kurzhals, "Stability and Control for the Manned Orbital Laboratory," presented to the SAE A-18 Committee Meeting, December 1963, available ntrs.nasa.gov/archive/nasa/casi.ntrs.nasa.gov (accessed 10 April 2009).
92 *National Aeronautics and Space Act of 1958*, Sec. 305, 2-i, available http://history.nasa.gov/spaceact.html (accessed 25 June 2008).
93 Erickson, op. cit., 58.

94 Eisenhower, op. cit., 257.
95 Michael J. Neufeld, "The End of the Army Space Program: Interservice Rivalry and the Transfer of the von Braun Group to NASA, 1958–1959," *Journal of Military History*, July 2005, vol. 69, no. 3, 739–40; "An Administrative History of NASA, 1958–1961," National Aeronautics and Space Administration, July 1965, 43–44, available http://history.nasa.gov/SP-4101.pdf
96 "Memorandum of Conference with the President: 1 December 1959," Eisenhower Library, Papers as President of the United States, 1953–61, DDE Diaries, Box 28.
97 Numerous scholarly articles in the years preceding the shuttle's first launch discuss its overt military design and intent; for example, Draper, Buck, and Goesch, "A Delta Shuttle Orbiter," 1971, 26–35. Written by engineers at the Air Force Flight Dynamics Laboratory, this article presents the argument for the shuttle's delta wing design, specifically to accommodate military missions. Robert Gillette, "Space Shuttle: A Giant Step for NASA and the Military?" *Science*, no. 171, 12 March 1971, 991–3, argues that that the military will be the shuttle's principal user and that its design must meet military needs. See also William G. Holder, "The Many Faces of the Space Shuttle," *Air University Review*, vol. 24, July/August 1973, 23–35.
98 Office of Science & Technology Policy, Executive Office of the President, 6 January 2005.
99 Burrows, *This New Ocean*, 1999, Chapter 10.
100 US Congress, House, Committee on Science and Astronautics, *Astronautical and Aeronautical Events of 1962: Report*, 88th Congress, 1st session, 12 June 1963, 17–18.
101 R. Cargill Hall, "Lunar Impact: The History of Project Ranger," National Aeronautics and Space Administration, 1977, available http://ntrs.nasa.gov/archive/nasa/casi.ntrs.nasa.gov/19780007206_1978007206.pdf (accessed 20 October 2007).
102 R. Cargill Hall, "Origins of US Space Policy: Eisenhower, Open Skies, and Freedom of Space," in John M. Logsdon (ed.), *Exploring the Unknown: Selected Documents in the History of the US Civil Space Program*, NASA SP-4407, Washington, DC: Government Printing Office, 1995, 215.
103 William B. Quandt, "The Electoral Cycle and the Conduct of Foreign Policy," *Political Science Quarterly*, vol. 101, no. 5, 1986, 834.
104 "Deterrence & Survival in the Nuclear Age," Security Resources Panel of the Science Advisory Committee, Executive Office of the President, 7 November 1957, 3, available www.gwu.edu/~nsarchiv/NSAEBB/NSAEBB139/nitze02.pdf (accessed 13 June 2007).
105 Roger D. Launius, "NASA Looks to the East: American Intelligence Estimates of Soviet Capabilities and Project Apollo," *Air Power History*, no. 3, Fall 2001, 7.
106 Erickson, op. cit., 283–4.
107 Marcia S. Smith, "US Space Programs: Civilian, Military, and Commercial," *CRS Issue Brief for Congress*, 28 September 2004, available www.nti.org/e_research/official_docs/other_us/crs092804.pdf (accessed 9 July 2009).
108 John F. Kennedy Presidential Recordings, Tape 111/A46, meeting with James Webb, 18 September 1963, John F. Kennedy Library, available www.jfklibrary.org/About-Us/News-and-Press/Press-Releases/JFK-Library-Releases-Recording-of–President-Kennedy-Discussing-Race-to-the-Moon.aspx (accessed 20 August 2009).
109 Ibid.
110 National Security Action Memorandum no. 271, 12 November 1963, available www.jfklibrary.org/Asset-Viewer/qVncp893wEmJFplIn1AlHA.aspx (accessed 20 August 2009).

111 Brian Harvey, *Soviet and Russian Lunar Exploration*, London: Praxis Publishing, 2006, 53; also interview with Sergei Khruschev, available www.pbs.org/redfiles/ (accessed 10 May 2009).

112 Frank Sietzen, "Soviets Planned to Accept JFK's Joint Lunar Mission Offer," *Space War*, 2 October 1997, available www.spacewar.com/news/russia-97h.html (accessed 12 June 2009).

113 House, Subcommittee on Manned Space Flight of the Committee on Science and Astronautics, 1974 NASA Authorization, Hearings on H.R. 4567, 93/2, Part 2, 1271.

114 S.N. Khrushchev, op. cit., 450; also published in *Department of State Bulletin*, 7 October 1963, 532–3.

115 L.T. Krug, op. cit., 36.

116 US Central Intelligence Agency, "National Intelligence Estimate 11–1–67: The Soviet Space Program," Washington, DC, 2 March 1967, 11 (declassified 1992).

117 Frank Borman and R.J. Sterling, *Countdown: An Autobiography*, New York: William Morrow, 1988, 189.

118 Oberg, *Space Power Theory*, 1999, 51.

119 Peter Pesavento and Charles Vick, *Quest*, vol. 11, no. 1, 2004, 17.

120 Christopher Andrew, *The World Was Going Our Way: The KGB and the Battle for the Third World*, New York: Basic Books, 2005, 38.

121 Sherman Kent, memo to Director of Central Intelligence Vice Admiral William Raborn, 1 January 1964.

122 Pesavento and Vick, op. cit., 98.

123 Siddiqi, *The Soviet Space Race with Apollo*, 2003, 745–54.

124 Speech by President Johnson at the United Nations Day Proclamation, 27 January 1967, *Public Papers of the Presidents of the United States: Lyndon B. Johnson, 1967*, vol. 1, Washington, DC: Government Printing Office, 1968, entry 18, pp. 91–2.

125 Henry Owen, "Space Goals After the Lunar Landing," US Department of State, October 1966.

126 Yuri Y. Karash, *The Superpower Odyssey: A Russian Perspective on Space Cooperation*, Herndon, VA: American Institute of Aeronautics & Astronautics, 1999, 76–9.

127 Herbert F. York, "Nuclear Deterrence and the Military Uses of Space," *Daedalus*, vol. 114, no. 2, *Weapons in Space, vol. I: Concepts and Technologies*, Spring 1985, 17–32.

128 Day, Logsdon, and Latell (eds.), *Eye in the Sky*, 1998, 69.

129 Paul B. Stares, *The Militarization of Space: US Policy, 1945–1984*, Ithaca, NY: Cornell University Press, 1985, 201–4.

130 Siddiqi, "The Soviet Co-Orbital Anti-Satellite System," 1997, 225.

131 Paul Stares, "US and Soviet Military Space Programs: A Comparative Assessment," *Daedalus*, vol. 114, no. 2, *Weapons in Space, vol. I: Concepts and Technologies*, Spring 1985, 133.

132 Peter Jankowitsch, "Legal Aspects of Military Space Activities," in Nandasiri Jasentuliyana (ed.), *Space Law: Development and Scope*, New York: Praeger Publishers, 1992, 150–1.

133 Matthew Mowthorpe, *The Militarization and Weaponization of Space*, New York: Lexington Books, 2003, Chapter 7, esp. 166–8.

134 R.D. Launius, "The View from Washington," *The History of the European Space Agency: Proceedings of an International Symposium*, Paris: ESA Publication Division, 1999.

135 Kazuto Suzuki, *Policy Logics and Institutions of European Space Collaboration*, Farnham, Surrey, UK: Ashgate Publishing, 2003, Chapter 2.

136 John Krige, "Building a Third Space Power," in Launius, Logsdon, and Smith (eds.), *Reconsidering Sputnik*, 2000, 290–91.

137 Jeremy Stocker, *Britain and Ballistic Missile Defence, 1942–2002*, London: Routledge, 2004, 36–8.

138 "The Atomic Bomb: Memorandum by the Prime Minister," GEN 75/1, 28 August 1945 (declassified), available http://extras.timesonline.co.uk/atomicbombcabinetoffice.pdf (accessed 23 May 2009).

139 Ian Beckett and John Gooch (eds.), *Politicians and Defence: Studies in the Formulation of British Defence Policy 1845–1970*, Manchester, UK: Manchester University Press, 1983, 133–4.

140 John Becklake, "German Engineers: Their Contribution to British Rocket Technology after World War II," in P. Jung (ed.), *History of Rocketry and Astronautics*, AAS History Series, vol. 22, *Proceedings of the 27th History Symposium of the International Academy of Astronautics*, Graz, 1993, Springfield, VA: American Astronautical Society, 1998, 157–72.

141 Stephen Robert Twigge, *The Early Development of Guided Missiles in the United Kingdom*, London: Routledge, 1993, 15.

142 Brian Harvey, *Europe's Space Programme: To Ariane and Beyond*, New York: Springer Publishers, 2009, 42.

143 Speech in Lyon, 28 September 1963, in Charles de Gaulle, *Discours & Messages*, Plon, vol. 4, 137.

144 Beatrice Heuser, *NATO, Britain, France and the FRG: Nuclear Strategies and Forces for Europe, 1949–2000*, New York: Macmillan Press, 1997, 66.

145 Colette Barbier, "The French Decision to Develop a Military Nuclear Programme in the 1950s," *Diplomacy & Statecraft*, vol. 4, no. 1, March 1993, 106.

146 Alain Gaubert, "Public Funding of Space Activities: A Case of Semantics and Misdirection," *Space Policy*, vol. 18, no. 4, November 2002, 287–92.

147 Statute of the European Space Agency, Article III.

148 Sheehan, *International Politics of Space*, 2007, 72–3.

149 "White Paper Space: A New European Frontier for an Expanding Union – An Action Plan for Implementing the European Space Policy," available http://europa.eu/documents/comm/white_papers/index_en.htm (accessed 20 June 2010).

150 Communication from the European Commission to the Council and European Parliament, "Space: A New European Frontier for an Expanding Union," COM (2003)673, 11 November 2003.

151 Matake Kamiya, "Nuclear Japan: Oxymoron or Coming Soon?," *The Washington Quarterly*, Winter 2002–2003, 72.

152 Japan Aerospace Exploration Agency, available www.isas.ac.jp/e/enterp/missions/index.shtml (accessed 9 January 2008).

153 National Space Development Agency of Japan, available http://warp.ndl.go.jp (accessed 9 January 2008).

154 Yuri Kase, "Japan's Nonnuclear Weapons Policy in the Changing Security Environment: Issues, Challenges, and Strategies," *World Affairs*, vol. 165, no. 3, Winter 2003, 126.

Chapter 3 First tier space actors: Launching BRICS into space

1 Dwight D. Eisenhower, address to the American Society of Newspaper Editors, Washington, DC, 16 April 1953.

2 Wade L. Huntley, "The Mice That Soar: Smaller States' Perspectives on Space Weaponization," in Bormann and Sheehan (eds.), *Securing Outer Space*, 2009, 147.

3 *Space Security 2008*, Project Ploughshares, University of Waterloo, 2008, 112, available www.spacesecurity.org (accessed 9 April 2009).

4 Klein, *Space Warfare*, 2006, 116.

5 Maldifassi and Abetti, *Defense Industries in Latin American Countries*, 1994, 2–3.
6 Trevor J. Pinch and Wiebe E. Bijker, "The Social Construction of Facts and Artifacts: Or How the Sociology of Science and the Sociology of Technology Might Benefit Each Other," *Social Studies of Science*, no. 14, August 1984, 399–441.
7 Joan Johnson-Freese, *Space as a Strategic Asset*, 2007, 47–9.
8 "Country Classification," World Bank, available http://data.worldbank.org/about/country-classifications (accessed 28 August 2009).
9 *The Space Report: 2009*, available www.thespacereport.org (accessed 10 January 2010).
10 Newberry, "Latin American Countries with Space Programs," 2003, 40–4.
11 Hu and Men, "Rising of Modern China," 2002, 1–2.
12 S.D. Sagan, "Why Do States Build Nuclear Weapons?" 1996–1997, 56–8, 63–5.
13 Turnball, *Siege Weapons of the Far East*, 2002, 17.
14 Antonellis and Murray, "China's Space Program," 2003, 652.
15 Cheng, "Evolution of China's Strategic Nuclear Weapons," 2006, 255.
16 Saunders, "China's Future in Space," 2005, 21.
17 Iris Chang, *Thread of the Silkworm*, 1995, Chapters 9–10.
18 Lewis and Hua, "China's Ballistic Missile Programs," 1992, 8.
19 Dean Cheng, "The Chinese Space Program: A Twenty-first Century Fleet in Being?", in Mulvenon and Yang (eds.), *A Poverty of Riches*, 2003, 31.
20 Chen, "China's Space Policy," 1991, 122.
21 Cheng, op. cit., 242.
22 Kulacki and Lewis, *A Place for One's Mat*, 2009, 5.
23 Ito and Shibata, "The Dilemma of Mao Tse-tung," 1968, 59–60; S.N. Khrushchev, *Nikita Khrushchev*, 2001, 295.
24 Harvey, *China's Space Program*, 2004, 30.
25 Hans M. Kristensen, et al., *Chinese Nuclear Forces and US Nuclear War Planning*, The Federation of American Scientists & The Natural Resources Defense Council, November 2006, 43; Shannon N. Kile, Vitaly Fedchenko, and Hans M. Kristensen, "World Nuclear Forces," *SIPRI Yearbook 2009: Armaments, Disarmament, and International Security*, Oxford, UK: Oxford University Press, 2009, 364–7.
26 Gilks, "China's Space Policy," 1997, 216.
27 Kulacki and Lewis, op. cit., 21.
28 Cunningham, "The Stellar Status Symbol," 2009, 80.
29 Chris Bulloch, "China's Satcoms: Relying on the West," Interavia Business & Technology, 9 November 2010.
30 F. Tillman Durdin, James Reston, and Seymour Topping, *New York Times Report from Red China*, Chicago: Quadrangle Books, 1971, 107.
31 Gilks, op. cit., 216.
32 Griffy-Brown and Pike, "Space Policy and Technology Management Strategies in China and Japan," 1994, 11.
33 Ross, *Negotiating Cooperation*, 1995, 159–60.
34 "China Eyes New Spaceport and Bigger Rockets", *Space Daily*, February 7, 2001, available www.spacedaily.com/news/china-01t.html (accessed 11 May 2007).
35 Victor Zaborsky, "Economics vs. Nonproliferation: US Launch Quota Policy toward Russia, Ukraine, and China," *The Nonproliferation Review*, Fall/Winter 2000, 153.
36 Michael N. Gold, "The Wrong Stuff: America's Aerospace Export Control Crisis," *Nebraska Law Review*, vol. 87, 2008, 521.
37 Dominique Moisi, "Europe Must Not Go the Way of Decadent Venice", *Financial Times*, London, 12 July 2005.
38 SCImago Research Group, 2010: www.scimagojr.com/countryrank.php.

39 Wayne C. Thompson and Steven W. Guerrier (eds.), *Space: National Programs and International Cooperation*, Boulder, CO: Westview Press, 1989, 93.
40 Gurtov, "Swords into Market Shares," 1993, 233–35; Chen, op. cit., 116–28.
41 Mora, "Civil-Military Relations in Cuba and China," 2002, 193.
42 Kan, "China and Proliferation of Weapons of Mass Destruction," 2009, 27; Davis, "China's Nonproliferation," 1995, 588–9.
43 Chinese Ministry of Science and Technology, available www.most.gov.cn (accessed 20 May 2008).
44 Chen, op. cit., 120.
45 He, "What Next for China in Space?" 2003, 185.
46 Xu, "The Launching of Aussat," 1992, 17–20.
47 Mary Amiti and Caroline Freund, "China's Export Boom," *Finance and Development: A Quarterly Magazine of the IMF*, September 2007, vol. 44, no. 3, 39.
48 Harding, *A Fragile Relationship*, 1992, 7.
49 Zaborsky, "Economics vs. Nonproliferation," 2000, 152–3.
50 http://english.peopledaily.com.cn (accessed 30 April 2011).
51 Johnson-Freese and Erickson, "The Emerging China-EU Space Partnership," 2006, 13.
52 Kevin Gallagher and Roberto Porzecanski, *The Dragon in the Room: China and the Future of Latin American Industrialization*, Palo Alto, CA: Stanford University Press, 2010, 14.
53 "China Asks Closer Space Industry Cooperation," *XinhuaNews Agency*, 29 August 2006, available www.china.org.cn/english/2006/Aug/179444.htm (accessed 17 October 2007).
54 Johnson-Freese, "Space *Wei Qi*," 2004, 123–4.
55 Chandler, "Confident China Joins Space Elite," 2003, 6.
56 Peter Aldhous and Anil Ananthaswamy, "Asia Blazes Trail to the Final Frontier," *New Scientist*, vol. 188, no. 2522, October 22–28, 2005, 8.
57 Paul Rincon, "What's Driving China Space Efforts?" BBC News, 25 September 2008.
58 Kulacki and Lewis, op. cit., 12–13.
59 Kathrin Hi, "China Plans Manned Space Station by 2020," *Financial Times*, 31 October 2010.
60 Johnson-Freese, "Strategic Communication with China," 2006, 43.
61 Chinese Academy of Sciences, "Chinese Space Station 'Tiangong I' Expected to Take Off in 2010," available http://english.cas.cn/Ne/CN/200911/t20091112_47093.shtml (accessed 22 November 2007).
62 ChinaNet News, 29 September 2009, available http://news.sina.com.cn/c/2008–09–29/145316381853.shtml (accessed 13 January 2010).
63 FUTRON, "China and the Second Space Age," 15 October 2003, 3, available www.futron.com/upload/wysiwyg/Resources/Whitepapers/China_n_%20Second_Space_Age_1003.pdf (accessed 13 November 2007).
64 "China Issues S&T Development Guidelines," Government of the People's Republic of China, 9 February 2006, available www.gov.cn/english/2006–02/09/content_183426.htm (accessed 13 November 2007).
65 Available www.sinodefence.com/space/facility/wenchang.asp (accessed 13 November 2007).
66 "China Asks Closer Space Industry Cooperation," Xinhua News Agency, 29 August 2006, available www.china.org.cn/english/2006/Aug/179444.htm (access 14 November 2007).
67 Saich, "The Fourteenth Party Congress," 1992, 1139.
68 Medeiros and Fravel, "China's New Diplomacy," 2003, 33.
69 Sheppard and DeVore, "China Rising," *China's Aerospace and Defence Industry*, 2000, section 10.1.

70 Ibid.

71 Wang, "China's Space Industry," 1996, 2–3.

72 Zhu and Xu, "Status and Prospects of China's Space Programme," 1997, 69–75.

73 "Jainbing 3 (Ziyuan 2) Imagery Reconnaissance Satellite," available www.sino-defence.com (accessed 9 June 2008).

74 "Homemade RDSS Satellite Systems for the First Times Exhibit Resolution up to 0.5m," available http://orig.news.ifeng.com/mil/2/200707/0718_340_158622.shtml (accessed 11 June 2008).

75 Hagt, "China's ASAT Test," 2007, 44.

76 Available http://english.peopledaily.com.cn/200610/12/eng20061012_311149.html (accessed 12 August 2007).

77 "China's Space Activities in 2006 (White Paper)," available www.china.org.cn/english/features/book/183672.htm (accessed 11 June 2007).

78 Ibid.

79 Ajay Lele, "China's ASAT Test: The Implications for India," Institute of Peace and Conflict Studies, New Delhi, India, 18 April 2007, available www.ipcs.org/article/china/chinas-asat-test-implications-for-india-2276.html (accessed 5 December 2007).

80 "China's Muscle Flex in Space," *International Herald Tribune*, 21 January 2007.

81 Dipankar Banerjee, "Indian Perspectives on Space Security," in John M. Logsdon and James Clay Moltz (eds.), *Collective Security in Space: Asian Perspectives*, Space Policy Institute, George Washington University, January 2008, available www.gwu.edu/~spi/assets/docs/Collective%20Security%20in%20Space%20-%20Asian%20Perspectives%20-%20January%202008.pdf (accessed 16 May 2009).

82 Bates Gill and Martin Kleiber, "China's Space Odyssey: What the Antisatellite Test Reveals About Decision-Making in Beijing," *Foreign Affairs*, vol. 86, no. 2, 2007, 2.

83 Sun Huixian, Dai Shuwu, Yang Jianfeng, Wu Ji, and Jiang Jingshan, "Scientific Objectives and Payloads of Chang'E-1 Lunar Satellite," *Journal of Earth System Science*, vol. 114, no. 6, 2005, 790.

84 Bradley Perrett and James R. Asker, "Person of the Year: Qian Xuesen," *Aviation Week and Space Technology*, 4 January 2008.

85 Unattributed, "China to Launch Second Lunar Probe Before End of 2011," 13 November 2008, available www.moondaily.com/reports/China_Reveals_Its_First_Full_Map_Of_Moon_Surface_999.html (accessed 13 November 2008).

86 Hitchens, "US–Sino Relations in Space," 2007, 15.

87 Tellis, "Punching the US Military's 'Soft Ribs'," 2007, 2.

88 France and Adams, "Chinese Threat to US Superiority," 2005, 20.

89 Mary C. Fitzgerald, "China's Military Modernization and its Impact on the United States and the Asia-Pacific," Hudson Institute, 30 March 2007, 3, available www.hudson.org/files/publications/07_03_29_30_fitzgerald_statement.pdf (accessed 20 August 2007).

90 Johnson-Freese, "China's Space Ambitions," 2007, 21.

91 China–US Dialogue on Space, "Budget," World Security Institute, available www.wsichina.org/space/program.cfm?programid=3&charid=1 (accessed 14 June 2009).

92 Tania Branigan, "China Could Make Moon Landing in 2025," *Guardian*, 20 September 2010; also quoted in Nicola Casarini, "Asian Space Tigers Bare Their Teeth," *Asian Times*, 9 November 2007, available www.atimes.com/atimes/China/IK09Ad02.html (accessed 7 July 2007).

93 Yongchun Zheng, Ziyuan Ouyang, Chunlai Li, Jianzhong Liu, and Yongliao Zou, "China's Lunar Exploration Program: Present and Future," *Planetary and Space Science*, vol. 56, no. 7, May 2008, 883.

94 Joel Achenbach, "Mars and Moon are Out of NASA's Reach for Now, Review Panel Says," *Washington Post*, 9 September 2009.

95 "Chinese 'Dragon' Aircraft Revealed for First Time," Sohu.com. http://mil. news.sohu.com/20080111/n254590647.shtml; Yu Liu and Bo Yang, "Space Launch Using Magnetic Levitation Propellant Launch Systems," *Beijing University of Aeronautics and Astronautics*, vol. 31, no. 1, 2005, 105–10.

96 Xin Dingding, "US Spacecraft Sparks Arms Race Concerns," *China Daily*, 24 April 2010, available www.chinadaily.com.cn/world/2010–04/24/content_9770149. htm (accessed 19 September 2010).

97 Gordon Fairclough, "China's Long March to the Moon," *The Wall Street Journal*, 23 October 2007, B1.

98 Angus Deaton and Jean Dreze, "Poverty and Inequality in India: A Re-examination," *Economic and Political Weekly*, 7 September 2002, 3729; "New Global Poverty Estimates: What It Means For India," *World Bank*, available www.worldbank.org.in

99 Mark Sappenfield, "India's Practical Space Program: 'We Have Liftoff!'," *Christian Science Monitor*, vol. 100, no. 231, 23 October 2008, 1.

100 Tiwari, *Wings of Fire*, 1999, 153.

101 Mistry, "India's Emerging Space Program," 1997, 162.

102 Subrato Paul, "Remote Sensing in Indian Agriculture," RMSI, January 2009, 32.

103 Monika Kannan, "Drought Mitigation in Rrajasthan (India) through Remote Sensing and GIS," paper presented at the annual meeting of the International Studies Association, Honolulu, HI, 5 March 2005; Harris, "Current Policy Issues in Remote Sensing," 2003, 295.

104 Sagar Kulkarni Thiruvananthapuram, "India Readying Weapon to Destroy Enemy Satellites: Saraswat," Press Trust of India, 3 January 2010.

105 Peter Ford, "What's Behind Asia's Moon Race?" *Christian Science Monitor*, 25 October 2007.

106 Baskaran, "Export Control Regimes," 2003, 213.

107 Atomic Energy Commission, Government of India, *Atomic Energy and Space Research: A Profile of the Decade 1970–1980*, July 1970.

108 Ibid., 5.

109 Talbott, "Dealing with the Bomb in South Asia," 1999, 112.

110 Cirincione, Wolfsthal, and Rajkumar, *Deadly Arsenals*, 2005, 223.

111 Michael Kraig, "The Indian Drive towards Weaponization: The Agni Missile Program," *Federation of Atomic Scientists*, available www.fas.org/nuke/guide/india/missile/agni-improvements.htm (accessed 20 June 2009).

112 Gupta, "The Indian Arms Industry", 1990, 850.

113 Perkovich, *India's Nuclear Bomb*, 2001, 327–8.

114 Mistry, op. cit., 159.

115 "State of the Satellite Industry," Satellite Industry Association, June 2005, available www.sia.org/industry_overview (accessed 3 March 2009).

116 "India Confirms Payloads for Chandrayaan-2 Lunar Mission," *RIA Novosti*, 8 October 2010, available http://en.rian.ru/science/20100830/160399917.html (accessed 11 October 2010).

117 Sheehan, *International Politics of Space*, 2007, 153–4.

118 Neelam Mathews, "Open for Business," *Aviation Week & Space Technology*, vol. 169, no. 23, 15 December 2008, 34.

119 Raj Chengappa, "India's Space Odyssey," *India Today*, 5 February 2007, 60–6.

120 "India Unveils Space Plane," *The Times of India*, 21 July 2001.

121 S.K. Bhan, S.K. Saha, L.M. Pande, and J. Prasad, "Use of Remote Sensing and GIS Technology in Sustainable Agricultural Management and Development: Indian Experience," Indian Institute of Remote Sensing, NRSA (N.D.), 2009,

available www.ces.iisc.ernet.in/energy/HC270799/LM/SUSLUP/Thema5/617/ 617.pdf (accessed 20 May 2010).

122 Soares de Lima and Hirst, "Brazil as an Intermediate State," 2006, 21; Office of the Prime Minister, "UK backs Brazil as Permanent Security Council Member," 27 March 2009, available www.number10.gov.uk/Page18798 (accessed 13 April 2010).

123 Hitchens, "US Space Policy," 2002, 16–17.

124 Brazilian Space Agency, *National Program of Space Activities: 2005–2014*, Ministério da Ciencia e Tecnologia, Brasília, 2005, 8.

125 Reis Pereira, *Política espacial brasileira e a trajetória do INPE (1961–2007)*, 2008, 101–5.

126 Edmund Jan Osmańczyk, *Encyclopedia of the United Nations*, London: Taylor & Francis, 2003, 1334.

127 Michael A. Morris, *The Strait of Magellan*, Leiden, Netherlands: Martinus Nijhoff Publishers, 1989, 134.

128 Michael Barletta, "The Military Nuclear Program in Brazil," 1997, 2.

129 Jones and McDonough, *Brazil: Tracking Nuclear Proliferation 1998*, Washington, DC, Carnegie Endowment for International Peace, 1998, 2, available www.carnegieendowment.org/files/tracking_brazil.pdf (accessed 16 January 2007).

130 Ministério da Ciência e Tecnologia, available www.inpe.br/institucional/historia.php (accessed 11 February 2007).

131 Bastos-Netto, "Dilemmas in Space Strategy," 2001, 120.

132 T. Krug, "Space Technology and Environmental Monitoring in Brazil," 1998, 655.

133 Judith Perera, "Brazil's Parallel Nuclear Industry," *New Scientist*, 17 September 1987, 39.

134 Pande, "Missile Technology Control Regime," 1999, 925–6.

135 Spector, *Nuclear Ambitions*, 1990, 221.

136 Bowen, "Brazil's Accession to the MTCR," 1996, 88.

137 *Veja*, 14 August 1991.

138 World Nuclear Organization, available www.world-nuclear.org/info/inf95. html (accessed 6 July 2007).

139 Maldifassi and Abetti, *Defense Industries in Latin American Countries: Argentina, Brazil, and Chile*, Westport, CT: Praeger Publishers, 1994, 28–9.

140 Ethan B. Kapstein, "The Brazilian Defense Industry and the International System," *Political Science Quarterly*, vol. 105, no. 4, 1990–1991, 587.

141 Zaborsky, "The Brazilian Export Control System," 2003, 124; United States Arms Control and Disarmament Agency, *World Military Expenditures and Arms Transfers, 1989*, US Government Printing Office, 1990, 81.

142 Da Silva, *Armas de Guerra do Brasil*, 1989; Zaborsky, op. cit., 124.

143 *Arms Control Today*, November 1995, 20.

144 International Atomic Energy Agency, available www.iaea.org/OurWork/SV/Invo/factsheet.html (accessed 18 November 2008).

145 Castilho Ceballos, "The Brazilian Space Program," 1995, 203.

146 *Manchete*, 13 May 1989, FBIS-LAT 16 June 1989.

147 Peter Bond, *Jane's Space Recognition Guide*, New York: Harper Collins, 2008, 130.

148 Chow, *Emerging National Space Launch Programs*, 1993, 47.

149 United States Congressional Record, 28 February 1996, E241.

150 Cristiane Gattaz Bueno, João Amato Neto, and Mauro Catharino, "A Dynamic Relationship Framework for Innovation: Implications for the Brazilian Aerospace Strategy Operations," *Transactions of the SDPS*, vol. 10, no. 3, September 2006, 11.

151 Petroni *et al.*, "Basic Strategic Orientation of Big Space Agencies," 2009, 49.
152 Ibid., 55–56.
153 I-Shih Chang, "Space Launch Vehicle Reliability," 2001, 34–5, available www. aero.org/publications/crosslink/winter2001/03.html (accessed 2 August 2007).
154 Darly Henriques da Silva, "Brazilian Participation in the International Space Station (ISS) Program", *Space Policy*, 2005, 56–7.
155 Yun Zhao, "2002 Space Cooperation Protocol", 2005, 213.
156 On the other hand, in January 2001 Brazil's government announced a US$40 billion plan to cover a large part of the rainforest with 10,000 kilometers of highways, as well as dams, power lines, mines, gas and oilfields, canals, ports, and logging zones. See Kolk, "From Conflict to Cooperation," 1998, 1482.
157 Douglas C. Morton, Ruth S. DeFries, Yosio E. Shimabukuro, Liana O. Anderson, Egidio Arai, Fernando del Bon Espirito-Santo, Ramon Freitas, and Jeff Morisette, "Cropland expansion changes deforestation dynamics in the southern Brazilian Amazon," *Proceedings of the National Academy of Sciences*, vol. 103, no. 39, 26 September 2006, 14637–41; Peter N. Spotts, "Satellite Images Reveal Amazon Forest Shrinking Faster," *Christian Science Monitor*, 21 October 2005.
158 Thomaz Guedes da Costa, "Brazil's SIVAM: As it Monitors The Amazon, Will it Fulfill its Human Security Promise?", Woodrow Wilson International Center for Scholars, *ECSP REPORT*, no. 7, 2001, 47–8.
159 Galanternick, "Lost in Space," 2002, 2.
160 Marcia Smith, "Military Space Programs: Issues Concerning DOD's SBIRS and STSS Programs," Congressional Research Service, 9 November 2005.
161 National Congress of Brazil. 2009 Federal Budget, Agência Espacial Brasileira, available www.camara.gov.br (accessed 6 March 2010).
162 Lisa Kubiske, US Chief of Mission, Brasília, confidential embassy cable, 9 February 2009, available www.politicaexterna.com/17831/wikileaks-eua-brasil-11-novos-documentos-alcntara-e-outros-temas (accessed 1 July 2011).
163 "Empresa ucraniana termina fabricação de plataforma para Alcântara," Agência Espacial Brasileira, Ministério da Ciência e Tecnologia, 17 October 2006.
164 "Brasil revela novo programa de foguetes," *Folha de São Paulo*, 26 October 2005, FBIS Document LAP20051026032002.
165 Available www.spacedaily.com/reports/Russia_Begins_Elbowing_ Ukraine_ Out_From_Brazil_ Space_Program_999.html (accessed 22 September 2008).
166 Agência Espacial Brasileira, *Programa Nacional de Atividades Espaciais: 2005–2014*, Ministério da Ciência e Tecnologia, 2005, 82.
167 Agência Brasil, 28 June 2006.
168 "Estratégia Nacional de Defesa," Brazilian Ministry of Defense, 17 December 2008, available www.exercito.gov.br/05notic/paineis/2008/12dez/img/defesa.pdf (accessed 30 October 2009).
169 "Brazil's Pursuit of a Nuclear Submarine Raises Proliferation Concerns," *WMD Highlights*, March 2008, available www.wmdinsights.org/I23/I23_LA1_BrazilPursuit.htm (accessed 11 September 2009).
170 Andrew Hurrell, "Lula's Brazil," 2008, 56.
171 "Star One President Says Rules Favor Non-Brazilian Companies," *Space New Business Report*, 23 September 2003.

Chapter 4 Second tier space actors

1 Arthur C. Clarke, *Profiles of the Future*, Clayton, Australia: Warner Books, 1962, 36.
2 Tarikhi, "Iran's Space Programme," 2009, 161.
3 History of Telecommunications Manufacturing Company of Iran, available www.tci.ir/userfiles/18–19.pdf (accessed 5 June 2010).

4 Shapir, "Iran's Efforts to Conquer Space," 2005.
5 Art Dula, "Private Sector Activities in Outer Space," *International Law and Practice*, vol. 19, no. 1, 1985, 164.
6 Sobhani, *The Pragmatic Entente*, 1989, 71; Theodore H. Moran, "Iranian Defense Expenditures and the Social Crisis," *International Security*, vol. 3, no. 3, Winter 1978–79, 180.
7 *World Armaments and Disarmament: SIPRI Yearbook.*, MIT Press, Stockholm International Peace Research Institute, 1980; Leslie M. Pryor, "Arms and the Shah," *Foreign Policy*, no. 31, Summer 1978, 56–7.
8 Amuzegar, "Nuclear Iran: Perils and Prospects," 2006, 92.
9 Venter, *Iran's Nuclear Option*, 2005, 156.
10 Bermudez, "Iran's Missile Development," 1994, 48.
11 Elaine Scoliono, "Documents Detail Israeli Missile Deal with the Shah," *New York Times*, 1 April 1986, A17; Parsi, *Treacherous Alliance*, 2007, 74–5.
12 Beit-Hallahmi, "The Israeli Connection," 1991, 419–22.
13 United States Arms Control and Disarmament Agency, *World Military Expenditures and Arms Transfers, 1981–1991*, 1992.
14 McNaugher, "Ballistic Missiles and Chemical Weapons in the Iran–Iraq War," 1990, 8.
15 Hildreth, "Iran's Ballistic Missile Programs," 2008, 4.
16 Blank, *Natural Allies?* 2005, 37–45.
17 Kass, "Iran's Space Program", 2006, 24.
18 Ben-David, "Iran Successfully Tests Shahab 3," 2003; Gruselle, "The Final Frontier," 2007, 55; Bill Gertz, "Missiles In Iran of Concern To State," *Washington Times*, 11 September 1997, 1, A14.
19 BBC News, "Tehran Aims for Satellite Launch," 5 January 2004, available http://news.bbc.co.uk/2/hi/middle_east/3370143.stm (accessed 7 December 2009).
20 "About the Iranian Space Agency," available www.isa.ir/index.php (accessed 30 July 2010).
21 504th Meeting of Committee on the Peaceful Uses of Outer Space (COPUOS), 11 June 2003, 5.
22 Kass, op. cit.
23 David Whitehouse, "One Small Step for Iran: Why NASA has Competition," *Independent* (UK), 9 February 2009.
24 BBC News, "Iran Declares Key Nuclear Advance," 11 April 2006, available http://news.bbc.co.uk/2/hi/middle_east/4900260.stm (accessed 22 March 2007).
25 Charles P. Vick, "North Korean, Iranian, and Pakistani Common Russian, Chinese Nuclear Weapons Heritage and Tests, What does it Reveal about the Missile-Borne Warhead Development Status?" Part 1, 20 March 2007, available www.globalsecurity.org/wmd/world/dprk/nuke-warhead-dev1.htm (accessed 6 June 2008); Uzi Rubin, "New Developments in Iran's Missile Capabilities: Implications Beyond the Middle East," *Jerusalem Issue Briefs*, vol. 9, no. 7, 25 August 2009, 62.
26 Uzi Rubin, "Yes, We Should Worry About Iran's Satellite," *Wall Street Journal*, 21 February 2009, A11.
27 Harvey, Smid, and Perard, *Emerging Space Powers*, 2010, 298.
28 Barbara Opall-Rome, "Iranian Sat Launch Triggers Concern, Kudos," *Defense News*, 9 February 2009; "New Iran Rocket Launch Site Shows N. Korea Links," *Agence France Presse*, 5 March 2010.
29 Mazol, "Persia in Space," 2009, 1
30 Wehling, "Russian Nuclear and Missile Exports to Iran," 1999, 138.
31 "Iran's Omid Satellite Launched Into Orbit," *Tehran Times*, 4 February 2009, available www.tehrantimes.com/index_View.asp?code=188488 (accessed 10 February 2009).

32 S. Stramondo, C. Bignami, M. Chini, N. Pierdicca, and A. Tertulliani, "Satellite Radar and Optical Remote Sensing for Earthquake Damage Detection: Results from Different Case Studies," *International Journal of Remote Sensing*, vol. 27, no. 2, May 2006, 4441.

33 Mostafa Jafari, Mehdi N. Fesharaki, and Peyman Akhavan, "Establishing an Integrated KM System in Iran Aerospace Industries Organization," *Journal of Knowledge Management*, vol. 11, no. 1, 2007, 128.

34 "Iran and Israel Plan New Satellites," United Press International, 2 February 2010.

35 Charles P. Vick, "Shahab-5/Kosar," GlobalSecurity.org, available www.globalsecurity.org/wmd/world/iran/shahab-5.htm, 2 March 2007 (accessed 15 June 2010.

36 "Iran Launches Spacecraft Carrying Animals," *Tehran Times*, 4 February 2010.

37 Kass, op. cit., 25.

38 Bill Gertz, "Iran's Missile Test Fails after Takeoff," *Washington Times*, 22 September 2000, A5.

39 Keith Crane, Rollie Lal, and Jeffrey Martini, *Iran's Political, Demographic, and Economic Vulnerabilities*, Santa Monica, CA: RAND Corporation, 2008, 22–8.

40 Haghshenass, "Iran's Asymmetric Naval Warfare," 2008, 14.

41 Warner D. Farr, "The Third Temple's Holy of Holies: Israel's Nuclear Weapons," *Counterproliferation Paper no. 2*, USAF Counterproliferation Center, Air War College, Maxwell Air Force Base, Alabama, available http://www.au.af.mil/au/awc/awcgate/cpc-pubs/farr.htm (accessed 10 June 2008); Avner Cohen, "The Last Nuclear Moment," *New York Times*, 6 October 2003.

42 E.L. Zorn, "Israel's Quest for Satellite Intelligence", *Studies in Intelligence*, Winter/Spring 2001, 33, available https://www.cia.gov/library/center-for-the-study-of-intelligence/csi-publications/csi-studies/studies/winter_spring01/article04.pdf (accessed 13 June 2009).

43 Ibid., 37.

44 Yossi Shain, Isaac Ben Israel, Deganit Paikowsky, Chaim Eshed, Reuven Pedatzur, "Israel in Space," *Tel Aviv Workshop for Science Technology and Security*, Harold Hartog School of Government and Policy, 14 April 2005, 7, available http://spirit.tau.ac.il/government/SpaceEnglish.pdf (accessed 20 June 2009).

45 Quoted in LLC Books (ed.), *Israeli Space Program: Israel Space Agency, Tauvex, Palmachim Airbase, Israeli Nano Satellite Association, Shavit 2*, Memphis, TN: General Books LLC, May 2010, 4.

46 "ISA International Relations," available www.most.gov.il/English/Units/Israel+Space+Agency/ISA+International+Relations.htm (accessed 20 August 2010).

47 Available www.iai.co.il/sip_storage/files/1/36281.pdf (accessed 20 August 2010).

48 *Israel High-Tech & Investment Report*, January 2001, 4.

49 Yaakov Katz, "Iran Delayed Satellite Launch," *Jerusalem Post*, 22 January 2008.

50 Israel Ministry of Foreign Affairs, "Israel in Space," 1 January 2003, available www.mfa.gov.il/mfa/mfaarchive/2000_2009/2003/1/focus%20on%20israel-%20israel%20in%20space (accessed 21 August 2010).

51 Avner Cohen, *The Worst-Kept Secret: Israel's Bargain with the Bomb*, New York: Columbia University Press, 2010, 25.

52 "Nuclear Diversion in the US? 13 Years of Contradiction and Confusion," Report of the Comptroller General of the United States, 18 December 1978, available www.irmep.org/ila/nukes/NUMEC/co1162251.pdf (accessed 11 June 2009).

53 Norris *et al.*, "Israeli Nuclear Forces, 2002," 2002, 73–5; Brower, "A Propensity for Conflict," 1997, 14–15.

54 Gottschalk, "South Africa's Space Program," 2010, 36.

55 Sasha Polakow-Suransky, *Unspoken Alliance: Israel's Secret Relationship with Apartheid South Africa*, New York: Knopf Doubleday Publishing, 2010, 4–6.

56 Albright, "South Africa and the Affordable Bomb," 1994, 42.
57 John Seiler, "South African Perspectives and Responses to External Pressures," *The Journal of Modern African Studies*, vol. 13, no. 3, 1975, 447–68.
58 Purkitt and Burgess, *South Africa's Weapons of Mass Destruction*, 2005, 51; Polakow-Suransky, op. cit., 44.
59 Carus, *Ballistic Missiles in the Third World*, 1990, 25.
60 Polakow-Suransky, op. cit., 131.
61 Pabian, "South Africa's Nuclear Weapons Program," 1995, 10–11; Albright and Gay, "A Flash from the Past," 1997, 15–17.
62 Stumpf, "South Africa: Nuclear Technology and Nonproliferation," 1993, 458; Pabian, op. cit., 12; United States National Security Council brief, 22 October 1979, available www.gwu.edu/~nsarchiv/NSAEBB/NSAEBB181/sa21.pdf (accessed 18 July 2009).
63 "South Africa: Nuclear Case Closed?", US Department of State memo, December 12, 1993, available www.gwu.edu/~nsarchiv/NSAEBB/NSAEBB181/sa34.pdf (accessed 2 August 2009).
64 *Estimates of National Expenditure 2011*, National Treasury Republic of South Africa, 23 February 2011, 6.
65 South African Government Gazette, 16 January 2009, available www.sabinet.co.za/prod_gazette.html
66 Gottschalk, op. cit., 35.
67 Reuters, 5 August 2009.
68 Reuters, "Despite Challenges, Scientific Research Improves in Post-Apartheid South Africa," 27 May 2010.
69 Linda Nordling, "Africa Analysis: Does Africa Need to be in Space?", *Science and Development Network*, 29 September 2010, available www.scidev.net/en/new-technologies/opinions/africa-analysis-does-africa-need-to-be-in-space-.html (accessed 5 December 2010).
70 "SA Companies Awarded Lucrative Airbus Contracts," 21 September 2010, available tradeinvest.co.za (accessed 5 December 2010).
71 Hamza, "Inside Saddam's Secret Nuclear Program," 1998, 26–7.
72 Carus and Bermudez, "Iraq's Al-Husayn Missile Programme," 1990, 55.
73 Nolan, *Trappings of Power*, 1991, 53; Central Intelligence Agency, Directorate of Intelligence, "Argentina: Condor Missile Program at a Critical Juncture," 1 August 1990, available www.foia.cia.gov/browse_docs.asp?doc_no=0001175499 (accessed 6 August 2009).
74 United Nations Monitoring, Verification and Inspection Commission report, 445.
75 Mark Phythian, "Britain and the Supergun," *Crime, Law, and Social Change*, vol. 19, no. 4, 1993, 354–6.
76 Said, "Missile Proliferation in the Middle East," 2001, 125.
77 Francona, *Ally to Adversary*, 1999, 36.
78 UNSCOM, "Report to the Security Council: Annex A, Status of Material Balances in the Missile Area," 25 January 1999, available www.fas.org/news/un/iraq/s/990125/index.html (accessed 18 June 2008).
79 "Comprehensive Report of the Special Advisor to the DCI on Iraq's WMD," Central Intelligence Agency, no. 2, 30 September 2004, 2.

Chapter 5 Third tier space actors

1 Cited in Steve Deger and Leslie Ann Gibson, *The Little Book of Positive Quotations*, Minneapolis, MN: Fairview Press, 2006, 122.
2 De León, *Historia de la actividad espacial en la Argentina*, 2008, 42.
3 Sánchez Peña, "Sounding Rockets in Argentina," 1969, 4.
4 Redick, "The Tlatelolco Regime and Nonproliferation," 1981, 103.

5 Shai Feldman, *Nuclear Weapons and Arms Control in the Middle East*, Cambridge, MA: MIT Press, 1996, 49, 62, 66.

6 Sánchez Peña, "Scientific experiences using Argentinean sounding rockets in Antarctica," 2000, 302–3.

7 Docampo, *Desarrollo de vectores espaciales y tecnología misilística en Argentina*, 1993, 13.

8 Santoro, *Operación Cóndor II*, 1992, 15.

9 Docampo, op. cit., 17–18.

10 J. Samuel Fitch, "The Decline of US Military Influence in Latin America," *Journal of Interamerican Studies and World Affairs*, vol. 35, no. 9, Summer 1993, 15–17; Timmerman, *The Death Lobby*, 1991, 154–7.

11 Barcelona and Villalonga, *Relaciones carnales*, 1992, 65; "El brigadier Crespo admitió que hizo ocho viajes al Líbano," *El Clarín*, 6 June 1998.

12 DiGiovanna, *From Defense to Development?* 2003, 107.

13 Barcelona and Villalonga, op. cit., 185–99.

14 Scheetz, "A Peace Dividend in South America?" 1996.

15 www.invap.com.ar/

16 *Space News*, 28 June 2010, available www.spacenews.com/contracts/100628-arianespace-lands-arsat1-contract.html.

17 "National Space Program: Argentina in Space, 1997–2008," available www.conae.gov.ar

18 Hulse, "China's Expansion and US Withdrawal," 2007, 18–19.

19 T. Pirard, *Air and Cosmos: Aviation Magazine International*, no. 1866, 22 November 2002, 36–7.

20 Mistry, *Containing Missile Proliferation*, 2005, 79.

21 Classified cable from US Embassy in Buenos Aires (#07BUENOSAIRES1793), 10 September 2007, available http://wikileaks.fi/cable/2007/09/07 BUENOSAIRES1793.html

22 Daniel Gallo, "Probaron en secreto un cohete argentino," *La Nación*, 5 August 2007.

23 Alejandro Castro, "Cálculos de modos naturales de vibración en placas con aplicación al cálculo de las frecuencias de resonancia del motor cohete Tronador II," in Alberto Cardona, Norberto Nigro, Victorio Sonzogni, and Mario Storti (eds.) *Mecánica Computacional*, vol. 25, November 2006.

24 Laura García, "Argentina planea tener lanzador de satélites en 2013," *Science and Development Network*, 23 August 2010, available www.scidev.net/en/news/argentina-plans-to-have-own-satellite-launcher-by-2013-.html

25 "Construyen un nuevo cohete en la misma base del Cóndor II", *Perfil*, 27 July 2008.

26 "Un lanzador de satélites se ensamblará en Punta Indio," *El Argentino*, 26 September 2011, available www.elargentino.com/Content.aspx?Id=159472

27 "Plan espacial nacional: Argentina en el espacio, 2004–2015," available www.conae.gov.ar/prensa/PlanEspacial2004–2015.pdf

28 Ibid.

29 "El padre de la física potosina," *Qüid*, Publication of the Faculty of the Autonomous University of San Luís Potosí, 16 March 2005, available http://galia.fc.uaslp.mx/~uragani/cam/quid/quid%2018.pdf (accessed 21 July 2009).

30 "El motor de reacción del cohete SCT-1," Secretaría de Comunicaciones y Transportes, no. 9, vol. 2, November/December 1960, 49–55, available (in translation) http://ntrs.nasa.gov/archive/nasa/casi.ntrs.nasa.gov/19710066187_1971066187.pdf (accessed 10 February 2010).

31 Frutkin and Griffin, "Space Activity in Latin America," 1968, 185.

32 William J. Carrington and Enrica Detragiache, "How Extensive is the Brain Drain?", *Finance & Development*, vol. 36, no. 2, June 1999, 49.

33 Peter B. de Selding, "Boeing Lands Three-satellite Deal with Mexico," *Space News*, 20 December 2010, available www.spacenews.com/satellite_telecom/101220-boeing-sat-deal-mexico.html (accessed 21 December 2011).

34 "Asociación de la Industria Aeroespacial Mexicana," *El Occidental*, 14 October 2007; "Anuncian proyectos Samsung y Bombardier," Government of Querétaro, Mexico, 15 May 2009.

35 "Mexico to Create its First Space Center on Yucatan Peninsula," *Rianovosti*, 22 Apr 2010, available http://en.rian.ru/world/20100422/158694450.html (accessed 15 May 2010).

36 *Gaceta Parlamentaria*, LXI Legislatura Cámara de Diputados, no. 2890-II, 15 November 2009, 53.

37 Wernher von Braun and Frederick I. Ordway III, *History of Rocketry and Space Travel*, New York: Thomas Y. Crowell Publishing, 1966, 23.

38 An excellent summary of Paulet's contribution is in Sara Madueño Paulet de Vásquez, "Pedro Paulet: Peruvian Space and Rocket Pioneer," *21st Century Science & Technology*, Winter 2001–2002, available www.21stcenturysciencetech.com/articles/winter01/paulet.html (accessed 19 October 2008).

39 Aaron Karp, "The Frantic Third World Quest for Ballistic Missiles," *Bulletin of the Atomic Scientists*, June 1988, 14; Aaron Karp, "Ballistic Missiles in the Third World," *International Security*, vol. 9, Winter 1984–85, 168.

40 CONIDA, www.conida.gob.pe/actividades.html

41 Natalya Ivanova, "Cooperate to Survive: Strategic Alliances in the Russian Aerospace Industry," in Slavo Radosevic and Bert Sadowski, *International Industrial Networks and Industrial Restructuring in Central and Eastern Europe*, Springer, 2004, 141.

42 "Peru: Prospects for Acquiring Advanced Weapon Systems," National Intelligence Council Memorandum, 10 July 1996, available www.foia.cia.gov/docs/DOC_0001042324/DOC_0001042324.pdf

43 http://docs.peru.justia.com/federales/leyes/28799-jul-19–2006.pdf

44 *People's Daily Online*, 28 December 2006. english.people.com.cn/200612/28/eng20061228_336500.html; Guillermo O. Descalzo, "Pedro Paulet Mostajo: Un Visionario Peruano," *Air & Space Power Journal*, Español, Tercer Trimestre 2007, available www.airpower.au.af.mil/apjinternational//apj-s/2007/3tri07/descalzo.html (accessed 23 April 2009).

45 J.M. Canales R, H. Bedon, and J. Estela, "First Steps to Establish a Small Satellite Program in Peru," *Aerospace Conference, 2010 IEEE*, 15 April 2010.

46 *La Política Espacial del Perú*, CONIDA, Peruvian Space Agency, November 2009, www.conida.gob.pe/prensa/PDF/politica_espacial.pdf (accessed 12 August 2010).

47 Ibid, 9.

48 "¿Qué es el programa espacial civil ecuatoriano?", Agencia Espacial Civil Ecuatoriana, available http://exa.ec/ (accessed 15 June 2010).

49 Available http://exa.ec/queyparaque.htm (accessed 15 June 2010).

50 Ronnie Nader and Gonzalo Naranjo, "Project Daedalus: The First Latin American Microgravity Research Plane," 60th International Astronautical Congress 2009, available www.exa.ec/trabajos/Project%20DAEDALUS%20-%20The%20First%20Latin-American%20Microgravity%20Research%20Plane.pdf (accessed 15 June 2010).

51 Juan José Taccone and Uziel Nogueira (eds.), *Andean Report no. 2*, Inter-American Development Bank, June 2005, 66–7.

52 Author's translation of www.abae.gob.ve/paginas/mision_vision.html (accessed 22 June 2010).

53 Kerry Dumbaugh and Mark Sullivan, "China's Growing Interest in Latin America," *CRS Report for Congress*, 20 April 2005, 3.

54 "Texto del acuerdo entre Urugay y Venezuela para el uso de la posición orbital 78° en el programa Venesat 1," republished in *El Espectador*, Montevideo, Uruguay, 19 December 2006.

55 "Chávez acusa a medios de ejercer 'terrorismo'," *El Universo*, 16 May 2009, available www.eluniverso.com/2009/05/16/1/1361/58E58491639F440080B8 CA2C7C67C6EE.html

56 *El Universal*, 10 January 2009.

57 B. Milton, "Venezuela: Small Satellite Manufacturing Plant to be Opened by 2012," *Latin Daily Financial News*, 2 November 2010, available www.latindailyfinancialnews.com/index.php/en/business/world/6875-venezuela-satellite-manufacturing-plant-to-be-opened-in-2012.html (accessed 15 February 2011).

58 Available www.cce.gov.co/web/guest/objetivos-programa-satelital (accessed 16 June 2011).

59 Available www.accionsocial.gov.co/documentos/207_vision2019.pdf (accessed 16 June 2011).

60 Cristina Pabón, "Gobierno boliviano crea agencia especial," Science and Development Network, 23 February 2010, available www.scidev.net/en/news/bolivian-government-creates-space-agency-.html (accessed 20 June 2011).

61 Eduardo García, "Bolivia, China Team Up on Communications Satellite," Reuters, 1 April 2010, available http://in.reuters.com/article/idINN0111911 620100401 (accessed 30 June 2011).

62 "Bolivia anuncia la creación de una agencia espacial," *El País*, Madrid, Spain, 6 October 2009.

63 "Copper-bottomed: Chile's Strange Way of Paying for Defence," *The Economist*, 7 February 2002.

64 Available www.agenciaespacial.cl/?page_id=76 (accessed 23 July 2011).

65 Available www.amsat.cl/english/Cesar/cesar1_eng.htm (accessed 23 July 2011).

66 Oscar Arenales Vergara, "Latin-America and Colombia Space Policy Approach to Future International Developments: Useful Applications of Space Technology and Cooperation," UNU-IAS Working Paper no. 113, February 2004, 16.

67 Deichmann and Wood, "GIS, GPS, and Remote Sensing," 2001.

68 Hamdy A. Ashour, "The Egyptian Space Program and its Role in the Sustainable Peaceful Development of Egypt, Middle East and Africa," United Nations Office for Outer Space Affairs, International Heliophysical Year, 2007, available www.oosa.unvienna.org/pdf/sap/2007/morocco/presentations/6–5.pdf (accessed 25 March 2010).

69 *National Space Policy*, 2000, available www.dawodu.com/space.pdf (accessed 10 April 2009).

70 Available www.nasrda.gov.ng (accessed 12 April 2009).

71 O. Kufoniyi and J. O. Akinyede, "Mainstreaming Geospatial Information for Sustainable National Development in Nigeria," Proceedings of the ISPRS Congress, Istanbul, Turkey, 2004, available www.cartesia.es/geodoc/isprs2004/comm4/papers/360.pdf (accessed 26 June 2009).

72 J. Paul Stephens and Martin Sweeting, "Low Cost Small Satellites in Coordinated Constellations for Sustainable Space Programmes for Developing Countries," 54th International Astronautical Congress, Bremen, Germany, 28 September 2003.

73 Unattributed, "Nigeria to Launch Two New Space Satellites in Three Years," *IT News Africa*, 12 May 2007, available www.itnewsafrica.com/?p=24 (accessed 8 August 2009).

74 R.A. Boroffice, "The Nigerian Space Programme: An Update," *African Skies*, no. 12, October 2008, 40.

75 *Algeria Today*, Embassy of the People's Republic of Algeria, vol. 2, no. 3, 12 November 2007, 5, available www.algeria-us.org/ALGERIA%20TODAY/VOL%20nov.pdf (accessed 11 October 2009).

76 Hichem Bouledjout, "Algeria 'On Track' to Launch Second Satellite System," *Science and Development Network*, 2 November 2007, available www.scidev.net/en/news/algeria-on-track-to-launch-second-satellite-syst.html (accessed 14 October 2009).

77 Briefing by Thomas Graham, Jr., "Military Use of Outer Space," Naval War College, 1 May 2003, available www.eisenhowerinstitute.org/programs/globalpartnerships/fos/newfrontier/grahambriefing.htm (accessed 6 August 2009).

78 "The 'Summit of Solutions' Opens on a High Note," *World Summit on the Information Society*, 16 November 2005, available www.itu.int/wsis/tunis/newsroom/highlights/16nov.html (accessed 13 May 2009).

79 "Utilization of Landsat Data for Ecological Studies of the Arid Zones of Tunisia: The Arzotu Experiment," *Canadian Symposium on Remote Sensing*, Quebec, Canada, 16–18 May 1977.

80 www.arabsat.com/Public/pdf/ArabSatBookEng.pdf

81 "Tunisia: A Country that Looks Ahead," *CNRS International Magazine*, available http://www2.cnrs.fr/en/1346.htm (accessed 22 May 2009).

82 Bermudez, "Ballistic Missile Development in the DPRK," 1999, 2.

83 Pinkston, "North and South Korean Space Development," 2006, 208–9.

84 Ji, "China and North Korea," 2001, 390.

85 Bermudez, op. cit., 10.

86 Lee, "North Korean Missiles," 2001, 90.

87 "North Korea's Nuclear Program, 2003," *Bulletin of the Atomic Scientists*, vol. 59, no. 2, March/April 2003, 76.

88 United States Arms Control and Disarmament Agency, "World Military Expenditures and Arms Transfers, 1969–1978," 3 May 1984.

89 Pinkston, op. cit., 210.

90 *Munju Choson*, 24 October 1998.

91 Federation of American Scientists, "North Korea's Taepodong and Unha Missiles," 21 September 2010, available www.fas.org/programs/ssp/nukes/nuclearweapons/Taepodong.html (accessed 18 November 2010).

92 Steven Lee Myers, "US Calls North Korean Rocket a Failed Satellite," *New York Times*, 15 September 1998.

93 Larry A. Niksch, "North Korea's Nuclear Weapons Program," *CRS Report for Congress*, 1 August 2006, 12.

94 "Iranian Missile Experts in N. Korea," *Agence France-Presse*, 29 March 2009.

95 William J. Broad, "North Korean Missile Launch was a Failure, Experts Say," *New York Times*, 5 April 2009; Steven A. Hildreth, "North Korean Ballistic Missile Threat to the United States," United States Congressional Research Service, 24 February 2009.

96 United States Geological Service, available http://earthquake.usgs.gov/earthquakes/recenteqsww/Quakes/us2009hbaf.php#details (accessed 12 December 2009).

97 *Izvestiya*, 24 June 1994.

98 Global Security, available www.globalsecurity.org/space/world/dprk/missile-developments.htm (accessed 22 September 2010).

99 "朝鲜宣布发展太空计划抗衡"西方强权" ("North Korea Announced the Development of Counter Space Program to the Western Powers"), *Rodong Sinmun*, 2 August 2009.

100 Peyrouse, "Russia and India Face Kazakhstan's Space Ambitions," 2008, 6.

101 "A New Kazakhstan in a New World: President Nazarbayev's Strategic Vision," *Kazakhstan's Echo*, Embassy of Kazakhstan in the United States, 2 March 2007, available http://prosites-kazakhembus.homestead.com/echo36.html (accessed 15 January 2009).

102 "Kazakhstan Will Not Lower Rent for Baikonur Space Center," *The Space Daily*, available www.spacedaily.com/reports/Kazakhstan_Will_Not_Lower_Rent_For_Baikonur_Space_Center.html (accessed 16 May 2009).

103 "Kazakhstan to Diversify Satellite Suppliers in Future," *Ria Novosti*, 2 February 2010, available http://en.rian.ru/exsoviet/20100202/157749127.html

104 The Project "Creation of the Earth Remote Sensing Space System of the Republic of Kazakhstan," www.gharysh.kz/en/article_258.html

105 "Astana Reconsiders Russian Use of Baikonur," *Eurasia Daily Monitor*, vol. 4, no. 117, 15 June 2007, available www.jamestown.org/single/?no_cache=1&tx_ttnews%5Btt_news%5D=32809 (accessed 30 June 2009).

106 *Eastern Europe, Russia, and Central Asia 2003*, Regional Surveys of the World, 3rd edition, London: Routledge, 2002, 113.

107 "Азербайджан запустит в космос первый спутник связи" ("Azerbaijan to Launch into Space the First Satellite Communications"), available www.iks-media.ru/news/2865181.html (accessed 22 December 2009).

108 Available http://hrc.az/eng/news/politics/1391-president-ilham-aliyev-confirms-state.html (accessed 14 June 2010).

109 Speech of the Chief of State Management of Radiofrequencies, Aflatun Mammadov, in the meeting connected with the subject "The Problems of Space," UNO, Vienna, 8 February 2010, available www.dri.az/view.php?lang=en&menu=0&id=1811 (accessed 15 May 2010).

110 "International Cooperation in the Peaceful Uses of Outer Space: Activities of Member States," UN Committee on the Peaceful Use of Outer Space (UNCO-PUOS), 23 November 2004, 2–3, available www.oosa.unvienna.org/pdf/reports/ac105/AC105_832E.pdf (accessed 13 October 2007).

111 Nguyen Khoa Son, "Recent Activities of Vietnam in Space Technology Applications," 13th Annual Asia-Pacific Regional Space Agency Forum, 2006, available www.aprsaf.org/data/aprsaf13_data/11_VN2006report-Prof.SON.pdf (accessed 20 June 2008).

112 "VINASAT-1 Bodes Well for Vietnam's Telecoms Development," *People's Army Newspaper*, 4 April 2008, available www.qdnd.vn/QDNDSite/en-US/75/72/182/155/160/60455/Default.aspx (accessed 2 November 2010).

113 Pham Auh Tuan, "Recent Development and Future of Space Technology in Vietnam," 14th Session of the Asia-Pacific Regional Space Agency Forum, 2007, available www.aprsaf.org/data/aprsaf14_data/day2/P12_SPACE%20TECHNOLOGY%20IN%20VIETNAM.pdf (accessed 11 August 2009).

114 "Vietnam to Build National Space Centre This Year," *Vietnam Business News*, 25 March 2010, available http://vietnambusiness.asia/vietnam-to-build-national-space-centre-this-year/ (accessed 22 January 2011).

115 Schofield and Storey, "Energy Security and Southeast Asia," *Harvard Asia Quarterly*, vol. 9, no. 4, Fall 2005, available www.asiaquarterly.com/content/view/160/40 (accessed 10 January 2007).

116 *Indonesia's Equatorial Orbit Twin Satellites for Space-based Safety Application in the Disaster Mitigation and Relief Effort*, available www.aprsaf.org/data/aprsaf16_data/D3–1130_AP16_CR_LAPAN.pdf (accessed 20 December 2010).

117 United Press International, 11 November 2010.

118 "Bangladesh to Launch $200–$300 Million Broadcasting Satellite," *Satellite Today*, 16 September 2010, available www.satellitetoday.com/st/headlines/Bangladesh-to-Launch-$200-$300-Million-Broadcasting-Satellite_35103.html (accessed 13 November 2010).

119 Interview with author on 5 March 2009, and www.suparco.gov.pk/pages/functions.asp

120 *Pakistan's Nuclear Programme and Imports*, The International Institute for Strategic Studies, 10 September 2007, available www.iiss.org/publications/ (accessed 30 June 2008).

121 Salim Mehmud, "Remote Sensing Applications to Earth Resources Survey in Pakistan," in F. Shahrokhi, N. Jasentuliyana, and N. Tarabzouni (eds.), *Space Commercialization: Satellite Technology*, Reston, VA: American Institute of Aeronautics & Astronautics, 1990.

122 Mehmud, "Pakistan's Space Programme," 1989, 220.

123 William Triplett II, "Gen. Xiong Pays a Visit to Pakistan," *Washington Times*, 19 March 2002, 17.

124 Richard H. Speier, Brian G. Chow, and S. Rae Starr, *Nonproliferation Sanctions*, Santa Monica, CA: RAND Corporation, 2005.

125 Corera, *Shopping for Bombs*, 2006, 89.

126 Jing-dong Yuan, "Assessing Chinese Nonproliferation Policy: Progress, Problems and Issues for the United States," James Martin Center for Nonproliferation Studies, 12 October 2001, available http://cns.miis.edu/other/jdtest.htm (accessed 12 February 2008).

127 Shah, "Democracy on Hold in Pakistan," 2002, 69.

128 Pervez Musharraf, *In the Line of Fire*, New York: The Free Press, 2006, 54.

129 "Musharraf Outlines Agenda for Expediting Pak's Space Program," *Pakistan News Service*, 3 February 2005, available www.paktribune.com/news/index.shtml?92655 (accessed 20 June 2009).

130 Office of the President of the Islamic Republic of Pakistan, available www.presidentofpakistan.gov.pk/PRPressReleaseDetail.aspx?nPRPressReleaseId=558&nYear=2007&nMonth=4 (accessed 12 March 2008).

131 Available www.suparco.gov.pk/pages/rsss.asp (accessed 12 March 2008).

132 Frédéric Achard and Christine Estreguil, "Forest Classification of Southeast Asia using NOAA AVHRR Data," *Remote Sensing of the Environment*, vol. 54, no. 3, December 1995, 201.

133 Speech by Y.B. Tan Sri Muhyiddin Yassin, Minister of International Trade and Industry Malaysia, 27 May 2008, available www.miti.gov.my/cms/contentPrint.jsp?id=com.tms.cms.article.Article_7b036772-c0a81573-2d952d95-ea53f710&paging=0 (accessed 7 November 2009).

134 Agence France-Presse, 23 August 2005.

135 "Malaysia Issues Guidebook for Muslims in Space," Reuters, 6 October 2007.

136 Bill Wu, Jeng-Shing Chern, Yen-Sen Chen, and An-Ming Wu, "Sub-Orbital Experiments Using Sounding Rockets in Taiwan," Presentation at Fourth Asian Space Conference 2008, Taipei, Taiwan, 1–3 October 2008, available http://www2.nspo.org.tw/ASC2008/4th%20Asian%20Space%20Conference%202008/oral/S12–10.pdf (accessed 8 November 2009).

137 NSPO, "ESEMS Program," available www.nspo.org.tw/2008e/projects/projectESEMS/intro.htm (accessed 8 November 2009).

138 *Space Master Plan for Thailand, 2004–2014*, available http://lerson.org/public/space/2005SpacePlan091E.pdf (accessed 3 December 2009).

139 Available www.gistda.or.th (accessed 3 December 2009).

140 Janie Hulse, "Argentina Urges Brazil to Promote South American Space Agency," *Diálogo Américas*, 9 September 2011, available www.dialogo-americas.com/en_GB/articles/rmisa/features/regional_news/2011/09/09/aa-south-american-space-agency (accessed 11 November 2011).

141 APRSAF, *Sentinel Asia*, available www.aprsaf.org/text/about.html (accessed 11 November 2011).

142 He Qizhi, "Organizing Space Cooperation in the Asia-Pacific Region," *Space Policy*, vol. 9, no. 3, August 1993, 209–12.
143 Gottschalk, "Africa's Active Participation in the Space Enterprise," *African Skies*, no. 12, October 2008, 26.

Conclusion: Space policy in developing countries

1 Euroconsult, *Government Space Markets, World Prospects to 2017*, available www.euroconsult-ec.com/research-reports/space-industry-reports/government-space-markets-38–24.html (accessed 9 October 2010).
2 NASA, "Report of the Space Task Group, 1969," available www.hq.nasa.gov/office/pao/History/taskgrp.html (accessed 10 October 2010).

Bibliography

Abidi, Rachid, "Satellite Remote Sensing in Aid of Development: The Tunis Declaration," *Space Policy*, vol. 19, no. 2, May 2003, 143–5.

Ahmed, Samina, "Pakistan's Nuclear Weapons Program," *International Security*, vol. 23, no. 4, Spring 1999, 178–204.

Albright, David, "South Africa and the Affordable Bomb," *Bulletin of the Atomic Scientists*, vol. 50, no. 4, July/August 1994, 37–47.

Albright, David and Gay, Corey, "A Flash from the Past," *Bulletin of the Atomic Scientists*, vol. 53, no. 6, November/December 1997, 15–17.

"Alternatives for Future U.S. Space-Launch Capabilities," Washington, DC: United States Congressional Budget Office, October 2006.

Amuzegar, Jahangir, "Nuclear Iran: Perils and Prospects," *Middle East Policy*, vol. 13, no. 2, Summer 2006, 90–112.

Antonellis, Robert and Murray, William S., "China's Space Program: The Dragon Eyes the Moon (and Us)," *Orbis*, Fall 2003, 645–52.

Bamford, James, *Body of Secrets: Anatomy of the Ultra-Secret National Security Agency*, New York: Doubleday, 2001.

Barcelona, Eduardo and Villalonga, Julio, *Relaciones carnales: La verdadera historia de la construcción y destrucción del misil Cóndor II*, Buenos Aires: Planeta, 1992.

Barletta, Michael, "The Military Nuclear Program in Brazil," Stanford, CA: Center for International Security and Arms Control, August 1997, available http://iis-db.stanford.edu/pubs/10340/barletta.pdf

Baskaran, Angathar, "Export Control Regimes and India's Space and Missile Programmes," *India Quarterly*, vol. 58, nos. 3–4, 2003, New Delhi: Indian Council of World Affairs, 205–42.

Bastos-Netto, Demetrio, "Dilemmas in Space Strategy for Regional Powers: A Brazilian Perspective," *Toward Fusion of Air and Space: Surveying Developments and Assessing Choices for Small and Middle Powers*, Santa Monica, CA: Rand Corporation, 2003, 119–31.

Beit-Hallahmi, Benjamin, "The Israeli Connection: Whom Israel Arms and Why," *International Journal of Middle East Studies*, vol. 23, no. 3, 1991, 419–22.

Ben-David, Alon, "Iran Successfully Tests Shahab 3," *Jane's Defence Weekly*, 9 July 2003, available http://jdw.janes.com/

Bermudez, Jr., Joseph S., "Iran's Missile Development," in William C. Potter and Harlan W. Jencks (eds.), *The International Missile Bazaar: The New Supplier's Network*, Boulder, CO: Westview Press, 1994.

Bermudez, Jr., Joseph S., "A History of Ballistic Missile Development in the DPRK," Occasional Paper no. 2, Monterey, CA: Center for Nonproliferation Studies, November 1999.

Blank, Stephen J., *Natural Allies? Regional Security in Asia and Prospects for Indo-American Strategic Cooperation*, Carlisle, PA: Army War College Strategic Studies Institute, 2005.

Bormann, Natalie and Sheehan, Michael (eds.), *Securing Outer Space*, New York: Routledge, 2009.

Bowen, Wyn Q., "Brazil's Accession to the MTCR," *The Nonproliferation Review*, Spring/Summer 1996, 86–91.

Brower, Kenneth S., "A Propensity for Conflict: Potential Scenarios and Outcomes of War in the Middle East," *Jane's Intelligence Review*, Special Report no. 14, February 1997, 14–15.

Brzezinski, Matthew, *Red Moon Rising: Sputnik and the Hidden Rivalries that Ignited the Space Age*, New York: Macmillan, 2008.

Burrows, William E., *This New Ocean: The Story of the First Space Age*, New York: Random House, 1999.

Carus, W. Seth, *Ballistic Missiles in the Third World: Threat and Response*, Westport, CT: Greenwood Publishing, 1990.

Carus, W. Seth and Bermudez, Jr., Joseph S., "Iraq's Al-Husayn Missile Programme, Part I," *Jane's Intelligence Review*, vol. 2, no. 5, 1 May 1990, 204–9.

Castilho Ceballos, Décio, "The Brazilian Space Program: A Selective Strategy for Space Development and Business," *Space Policy*, vol. 11, no. 3, August 1995, 202–4.

Cervino, M., Corradini, S., and Davolio, S., "Is the 'Peaceful Use' of Outer Space Being Ruled Out?" *Space Policy*, vol. 19, no. 4, November 2003, 231–7.

Chandler, David L., "Confident China Joins Space Elite," *New Scientist*, vol. 180, no. 2418, 25–31 October 2003, 6–8.

Chang, Iris, *Thread of the Silkworm*, New York: Basic Books, 1995.

Chang, I-Shih, "Space Launch Vehicle Reliability," *Crosslink: The Aerospace Corporation Magazine of Advances in Aerospace Technology*, vol. 6, no. 2, Winter 2001, 23–32.

Chen, Yanping, "China's Space Policy: A Historical Review," *Space Policy*, vol. 7, no. 2, May 1991, 116–28.

Cheng, Ta-Chen, "The Evolution of China's Strategic Nuclear Weapons," *Defense & Security Analysis*, vol. 22, no. 3, 2006, 241–60.

Chow, Brian, *Emerging National Space Launch Programs: Economics and Safeguards*, Santa Monica, CA: Rand Corporation, 1993.

Cirincione, Joseph, Wolfsthal, Jon B., and Rajkumar, Miriam, *Deadly Arsenals*, Washington, DC: Carnegie Endowment for International Peace, June 2005.

Civilian Space Policy and Applications, Washington, DC: U.S. Government Printing Office, June 1982 [report by the Office of Technology Assessment].

Clegg, Elizabeth and Sheehan, Michael, "Space as an Engine of Development: India's Space Programme," *Contemporary South Asia*, vol. 3, no. 1, 1994, 25–35.

Corera, Gordon, *Shopping for Bombs: Nuclear Proliferation, Global Insecurity, and the Rise and Fall of the A.Q. Khan Network*, New York: Oxford University Press USA, 2006.

Cunningham, Fiona, "The Stellar Status Symbol: True Motives for China's Manned Space Program," *China Security*, vol. 5, no. 3, 2009, 73–88.

Da Silva, Reginaldo, *Armas de Guerra do Brasil*, São Paulo: Editora Nova Cultural Ltda., 1989.

Davis, Zachary S., "China's Nonproliferation and Export Control Policies," *Asian Survey*, vol. 35, no. 6, June 1995, 587–603.

Day, Dwayne A., Logsdon, John M., and Latell, Brian, *Eye in the Sky: The Story of the Corona Spy Satellites*, Washington, DC: Smithsonian Institution Press, 1998.

Deichmann, Uwe and Wood, Stanley, "GIS, GPS, and Remote Sensing," *Brief 7*, International Food Policy Research Institute, August 2001, available www.iapad.org/publications/ppgis/gis_gps_and_remote_sensing.pdf (accessed 30 August 2011).

De León, Pablo, *Historia de la actividad espacial en la Argentina*, Raleigh, NC: Lulu, 2008.

Dickson, Paul, *Sputnik: The Shock of the Century*, New York: Walker & Company, 2001.

DiGiovanna, Sean, *From Defense to Development? International Perspectives on Realizing the Peace Dividend*, New York: Routledge, 2003.

Divine, Robert A., *The Sputnik Challenge*, New York: Oxford University Press USA, 1993.

Docampo, César, *Desarrollo de vectores espaciales y tecnología misilística en Argentina: El Cóndor II*, Buenos Aires: EURAL, 1993.

Dolman, Everett C., *Astropolitik: Classic Geopolitics in the Space Age*, London: Frank Cass, 2001.

Dos Santos, Ilvaro Fabricio, "Financing of Space Assets," *Space Policy*, vol. 19, no. 2, May 2003, 127–9.

Draper, Alfred C., Buck, Melvin L., and Goesch, William H., "A Delta Shuttle Orbiter," *Astronautics & Aeronautics*, vol. 9, January 1971, 26–35.

Durant III, Frederick C. (ed.), *Between Sputnik and the Shuttle: New Perspectives on American Astronautics*, American Astronautical Society History Series, vol. 3, San Diego, CA: Univelt, Inc., 1981.

Eisenhower, Dwight D., *Waging Peace: 1956–1961*, New York: Doubleday & Company, 1965.

Elhefnawy, Nader, "The Rise and Fall of Great Space Powers," *The Space Review: Essays and Comentary about the Final Frontier*, 27 August 2007, available www.thespacereview.com/article/942/1

Erickson, Mark, *Into the Unknown Together: The DoD, NASA, and Early Spaceflight*, Montgomery, AL: Air University Press, 2005.

Euroconsult, *Government Space Markets: Forecasts to 2017*, 11th edn., Paris, France, 2008, available www.euroconsult-ec.com/research-reports/space-industry-reports/government-space-markets-38–24.html (accessed 9 October 2010).

European Policy Institute, *Yearbook on Space Policy, 2006/2007: New Impetus for Europe*, New York: Springer Publishers, 2008.

Figliola, Patricia Moloney, Behrens, Carl E., and Morgan, Daniel, "U.S. Space Programs: Civilian, Military, and Commercial," Issue brief IB92011, Washington, DC: Congressional Research Service, 13 June 2006.

Firth, Noel E. and Noren, James H., *Soviet Defense Spending: A History of CIA Estimates, 1950–1990*. College Station, TX: Texas A&M Press, 1998.

Flattau, Pamela Ebert, Bracken, Jerome, Van Atta, Richard H., De la Cruz, Rodolfo, and Sullivan, Kay, "The National Defense Education Act of 1958: Selected Outcomes," Washington, DC: Institute for Defense Analyses, Science & Technology Institute, March 2006.

Florini, Ann M. and Dehqanzada, Yahya A., "No More Secrets? Policy Implications of Commercial Remote Sensing Satellites," Washington, DC: Carnegie Endowment for International Peace, Carnegie Paper no. 1, July 1999.

France, Martin E. and Adams, Richard J., "The Chinese Threat to US Superiority," *High Frontier Journal.* vol. 1, no. 3, Winter 2005, 17–23.

Francona, Rick, *Ally to Adversary: An Eyewitness Account of Iraq's Fall from Grace.* Baltimore, MD: US Naval Institute Press, 1999.

Frutkin, Arnold W. and Griffin, Jr., Richard B., "Space Activity in Latin America," *Journal of Interamerican Studies*, vol. 10, no. 2, 1968, 185–93.

Gaggero, Eduardo D., "New Roles in Space for the 21st Century: A Uruguayan View," *Space Policy*, vol. 19, no. 3, August 2003, 203–10.

Galanternick, Mery, "Lost in Space: A Military Vision of Brazil in Space Finds Itself Grounded by Budget Realities," *Latin Trade*, November 2002.

George, Hubert, "Developing Countries and Remote Sensing: How Intergovernmental Factors Impede Progress," *Space Policy*, vol. 16, no. 4, November 2000, 267–73.

Geraci, Richard V., "Army Transformation War Game: Insights Concerning Space Operations," *The Army Space Journal*, vol. 2, no. 1, Winter/Spring 2003, 4–5.

Gilks, Anne, "China's Space Policy," *Space Policy*, vol. 13, no. 3, August 1997, 215–27.

Gottschalk, Keith, "The Roles of Africa's Institutions in Ensuring Africa's Active Participation in the Space Enterprise: The Case for an African Space Agency (ASA)," *African Skies*, no. 12, October 2008, 20–2.

Gottschalk, Keith, "South Africa's Space Program," *Astropolitics*, vol. 8, no. 1, January 2010, 35–48.

Gray, Colin S. and Sloan, Geoffrey R. (eds.), *Geopolitics, Geography and Strategy*, London: Frank Cass, 1999.

Griffy-Brown, Charla and Pike, Gordon, "An Analysis of Space Policy and Technology Management Strategies in China and Japan," Satellite Communications: the Ninth National Space Engineering Symposium Proceedings, 1994, 7–15.

Grimmett, Richard F., "Trends in Conventional Arms Transfers to the Third World by Major Suppliers, 1980–1987," Washington, DC: Congressional Research Service, 9 May 1988.

Gruselle, Bruno, "The Final Frontier: Missile Defence in Space?," *Disarmament Forum*, no. 1, 2007, 53–7.

Gupta, Amit, "The Indian Arms Industry: A Lumbering Giant?," *Asian Survey*, vol. 30, no. 9, September 1990, 846–61.

Gurtov, Mel, "Swords into Market Shares: China's Conversion of Military Industry to Civilian Production," *The China Quarterly*, no. 134, June 1993, 213–41.

Haghshenass, Fariborz, "Iran's Asymmetric Naval Warfare," *Policy Focus #87*, Washington Institute for Near East Policy, September 2008, available www.washingtoninstitute.org/pubPDFs/PolicyFocus87.pdf (accessed 3 March 2010).

Hagt, Eric, "China's ASAT Test: Strategic Response," *China Security*, no. 5, Winter 2007, 31–51.

Hamza, Khidhir, "Inside Saddam's Secret Nuclear Program," *Bulletin of the Atomic Scientists*, vol. 54, no. 5, September/October 1998, 26–33.

Handberg, Roger and Johnson-Freese, Joan, "The Return of the American Military to Crewed Spaceflight: Hypersonic and Other Visions," *Space Policy*, vol. 13, no. 4, November 1997, 295–304.

Hardesty, Von and Eisman, Gene, *Epic Rivalry: The Inside Story of the Soviet and American Space Race*, Washington, DC: National Geographic Books, 2007.

Harding, Harry, *A Fragile Relationship: The United States and China since 1972*, Washington, DC: Brookings Institution Press, 1992.

Harris, Ray, "Current Policy Issues in Remote Sensing: Report by the International Policy Advisory Committee of ISPRS," *Space Policy*, vol. 19, no. 4, 2003, 293–6.

Harvey, Brian, *China's Space Program: From Conception to Manned Spaceflight*, New York: Springer Publishing, 2004.

Harvey, Brian, Smid, Henk H.F. and Pirard, Theo, *Emerging Space Powers: The New Space Programs of Asia, the Middle East, and South America*, New York: Springer Publishing, 2010.

Hays, Peter L. and Lutes, Charles D., "Towards a Theory of Spacepower," *Space Policy*, vol. 23, no. 4, November 2007, 206–9.

He, Qizhi, "Organizing Space Cooperation in the Asia-Pacific Region," *Space Policy*, vol. 9, no. 3, August 1993, 209–12.

He, Sibing, "What Next for China in Space after Shenzhou?" *Space Policy*, vol. 19, no. 3, August 2003, 183–9.

Henriques da Silva, Darly, "Brazilian Participation in the International Space Station (ISS) Program: Commitment or Bargain Struck?" *Space Policy*, vol. 21, no. 1, February 2005, 55–63.

Hildreth, Steven A., "Iran's Ballistic Missile Programs: An Overview," Washington, DC: United States Congressional Research Service, 21 July 2008.

Hitchens, Theresa, "US Space Policy: Time to Stop and Think," *Disarmament Diplomacy*, no. 67, October–November 2002, 15–32.

Hitchens, Theresa, "U.S.–Sino Relations in Space: From 'War of Words' to Cold War in Space?," *China Security*, Winter 2007, 12–30, available www.wsichina.org/cs5_2.pdf (accessed 2 June 2010).

Hu, Angang and Men, Honghua, "The Rising of Modern China: Comprehensive National Power and Grand Strategy," *Strategy & Management*, no. 3, 2002.

Hulse, Janie, "China's Expansion into and U.S. Withdrawal from Argentina's Telecommunications and Space Industries and the Implications for U.S. National Security," Carlisle, PA: U.S. Army War College Strategic Studies Institute, September 2007, available www.strategicstudiesinstitute.army.mil/pdffiles/pub806.pdf

Hunt, Linda, *Secret Agenda: The United States Government, Nazi Scientists, and Project Paperclip, 1945 to 1990*, New York: St. Martin's Press, 1991.

Hurrell, Andrew, "Lula's Brazil: A Rising Power, but Going Where?" *Current History*, February 2008, 51–57.

Ito, Kikuzo and Shibata, Minoru, "The Dilemma of Mao Tse-tung," *The China Quarterly*, no. 35, July–September 1968, 58–77.

Jane's China's Aerospace and Defence Industry, London: Jane's Information Group, December 2000.

Ji, You, "China and North Korea: A Fragile Relationship of Strategic Convenience," *Journal of Contemporary China*, vol. 10, no. 28, August 2001, 387–98.

Jolly, Claire and Razi, Gohar, *The Space Economy at a Glance*, OECD International Futures Programme, Organisation for Economic Co-operation and Development, 2007.

Johnson-Freese, Joan, *The Chinese Space Program: A Mystery within a Maze*, Malabar, FL: Krieger Publishing, 1998.

Johnson-Freese, Joan, "Space *Wei Qi*: The Launch of *Shenzhou V*," *Naval War College Review*, vol. 57, no. 2, Spring 2004, 121–45.

Johnson-Freese, Joan, "Strategic Communication with China: What Message about Space?," *China Security*, no. 2, 2006, 37–57.

Johnson-Freese, Joan, *Space as a Strategic Asset*, New York: Columbia University Press, 2007.

Johnson-Freese, Joan, *China's Space Ambitions*, Proliferation Papers no. 18, Paris: Institut Français des Relations Internationales, 2007.

Johnson-Freese, Joan and Erickson, Andrew S., "The Emerging China–EU Space Partnership: A Geotechnological Balancer," *Space Policy*, vol. 22, no. 1, February 2006, 12–22.

Jones, Rodney W. and McDonough, Mark G., *Brazil: Tracking Nuclear Proliferation 1998*, Washington, DC: Carnegie Endowment for International Peace, 1998.

Kan, Shirley A., "China and Proliferation of Weapons of Mass Destruction and Missiles: Policy Issues," Washington, DC: Congressional Research Service, 27 July 2009.

Kass, Lee, "Iran's Space Program: The Next Genie in the Bottle?," *The Middle East Review of International Affairs*, vol. 10, no. 3, September 2006, 15–32, available http://meria.idc.ac.il/journal/2006/issue3/jv10no3a2.html

Kay, W.D., *Can Democracies Fly in Space? The Challenge of Revitalizing the U.S. Space Program*, Westport, CT: Praeger Publishing, 1995.

Kennedy, John F., Presidential Recordings, tape 111/A46, meeting with James Webb, 18 September 1963. John F. Kennedy Library, available http://history.nasa.gov/JFK-Webbconv/pages/backgnd.html#interest

Khrushchev, Sergei N., *Nikita Khrushchev and the Creation of a Superpower*, University Park, PA: Pennsylvania State University Press, 2001.

Klein, John J., *Space Warfare: Strategy, Principles, and Policy*, New York: Routledge, 2006.

Kolk, Ans, "From Conflict to Cooperation: International Policies to Protect the Brazilian Amazon," *World Development*, vol. 26, no. 8, August 1998, 1481–93.

Krug, Linda T., *Presidential Perspectives on Space Exploration: Guiding Metaphors from Eisenhower to Bush*, New York: Praeger Publishing, 1991.

Krug, Thelma, "Space Technology and Environmental Monitoring in Brazil," *Journal of International Affairs*, vol. 51, no. 2, Spring 1998, 655–74.

Kulacki, Gregory and Lewis, Jeffrey G., *A Place for One's Mat: China's Space Program, 1956–2003*, Cambridge, MA: American Academy of Arts and Sciences, 2009.

Lasby, Clarence G., *Project Paperclip: German Scientists and the Cold War*, New York: Athenaeum, 2nd edn., 1975.

Launius, Roger D., "Eisenhower, Sputnik, and the Creation of NASA," *Prologue: The Quarterly Journal of the National Archives and Record Administration*, Summer 1996, 127–40.

Launius, Roger D., Logsdon, John M., and Smith, Robert W. (eds.), *Reconsidering Sputnik: Forty Years Since the Soviet Satellite*, New York: Routledge, 2000.

Lee, Chung Min, "North Korean Missiles: Strategic Implications and Policy Responses," *The Pacific Review*, vol. 14, no. 1, March 2001, 85–120.

Lewis, John Wilson and Hua, Di, "China's Ballistic Missile Programs: Technologies, Strategies, Goals," *International Security*, vol. 17, no. 2, Fall 1992, 5–40.

Logsdon, John M. (ed.), *Exploring the Unknown: Selected Documents in the History of the U.S. Civil Space Program*, vol. 1, Washington, DC: Government Printing Office, 1995–2004.

Maldifassi, José O. and Abetti, Pier A., *Defense Industries in Latin American Countries*, Westport, CT: Greenwood Publishing, 1994.

Mazol, James, "Persia in Space: Implications for U.S. National Security," *Policy Outlook*, Arlington, VA: George C. Marshall Institute, February 2009, 1–4.

McDougall, Walter A., *The Heavens and the Earth: A Political History of the Space Age*, New York: Basic Books, 1985.

McNaugher, Thomas L., "Ballistic Missiles and Chemical Weapons in the Iran–Iraq War," *International Security*, vol. 15, no. 2, Fall 1990, 5–34.

Medeiros, Evan S. and Fravel, Taylor M., "China's New Diplomacy," *Foreign Affairs*, vol. 82, no. 6, November/December 2003, 22–35.

Mehmud, Salim, "Pakistan's Space Programme," *Space Policy*, vol. 5, no. 3, August 1989, 217–26.

Mistry, Dinshaw, "India's Emerging Space Program," *Pacific Affairs*, vol. 71, no. 2, Summer 1998, 151–93.

Mistry, Dinshaw, *Containing Missile Proliferation: Strategic Technology, Security Regimes, and International Cooperation in Arms Control*, Seattle: University of Washington Press, 2005.

Moltz, James Clay, "Protecting Safe Access to Space: Lessons from the First 50 Years of Space Security," *Space Policy*, vol. 23, no. 4, November 2007, 199–205.

Mora, Frank O., "A Comparative Study of Civil–Military Relations in Cuba and China: The Effects of Bingshang," *Armed Forces & Society*, vol. 28, no. 2, Winter 2002, 185–209.

Mulvenon, James C. and Yang, Andrew N.D. (eds.), *A Poverty of Riches: New Challenges and Opportunities in PLA Research*, Santa Monica, CA: RAND Corporation, 2003.

Programa Nacional de Atividades Espaciais: 2005–2014 [National Program of Space Activities: 2005–2014], Brasília: Agência Espacial Brasileira/Ministério da Ciência e Tecnologia, 2005.

Neal, Homer A., Smith, Tobin L., and McCormick, Jennifer B., *Beyond Sputnik: U.S. Science Policy in the Twenty-First Century*. Ann Arbor, MI: University of Michigan Press, 2008.

Neufeld, Michael J., *Von Braun: Dreamer of Space, Engineer of War*, New York: Knopf Publishing, 2007.

Newberry, Robert D., "Latin American Countries with Space Programs: Colleagues or Competitors?" *Air & Space Power Journal*, vol. 17, no. 3, Fall 2003, 39–48.

Nolan, Janne E., *Trappings of Power: Ballistic Missiles in the Third World*, Washington, DC: Brookings Institution Press, March 1991.

Norris, Robert S., Arkin, William, Kristensen, Hans M., and Handler, Joshua, "Israeli Nuclear Forces, 2002," *Bulletin of the Atomic Scientists*, vol. 58, no. 5, September/October 2002, 73–5.

Norris, Robert S., Kristensen, Hans M., and Handler, Joshua, "North Korea's Nuclear Program, 2003," NRDC Nuclear Notebook, *Bulletin of the Atomic Scientists*, vol. 59, no. 2, March/April 2003, 74–7.

Nye, Jr., Joseph S., *The Paradox of American Power: Why the World's Only Superpower Can't Go It Alone*, New York: Oxford University Press, 2002.

Oberg, Jim, *Space Power Theory*, Maxwell Air Force Base, AL: Air University Press, 1999.

Ordway, Frederick I., Sharpe, Mitchell R., and Wakeford, Ronald C., "Project Horizon: An Early Study of a Lunar Outpost," *Acta Astronautica*, vol. 17, no. 10, 1988, 1105–21.

Pabian, Frank V., "South Africa's Nuclear Weapons Program: Lessons for U.S. Nonproliferation Policy," *The Nonproliferation Review*, vol. 3, no. 1, Fall 1995, 1–19.

Pande, Savita, "Missile Technology Control Regime: Impact Assessment," *Strategic Analysis: A Monthly Journal of the IDSA*, vol. 23, no. 6, September 1999, 925–40.

Parsi, Trita, *Treacherous Alliance: The Secret Dealings of Israel, Iran, and the U.S.*, New Haven, CT: Yale University Press, 2007.

Partington, James Riddick, *A History of Greek Fire and Gunpowder*, Baltimore, MD: Johns Hopkins University Press, 1999.

Perkovich, George, *India's Nuclear Bomb: The Impact on Global Proliferation*, Berkeley, CA: University of California Press, 2001.

Petroni, Giorgio, Venturini, Karen, Verbano, Chiara, and Cantarello, Silvia, "Discovering the Basic Strategic Orientation of Big Space Agencies," *Space Policy*, vol. 25, no. 1, February 2009, 45–62.

Peyrouse, Sébastien, "Russia and India Face Kazakhstan's Space Ambitions," *Central Asia-Caucasus Analyst*, no. 6, 29 October 2008, 6–8, available www.cacianalyst.org/?q=node/4971 (accessed 1 June 2010).

Pinkston, Daniel A., "North and South Korean Space Development: Prospects for Cooperation and Conflict," *Astropolitics*, vol. 4, no. 2, August 2006, 207–27.

Purkitt, Helen E. and Burgess, Stephen F., *South Africa's Weapons of Mass Destruction*, Bloomington, IN: Indiana University Press, 2005.

Redick, John R., "The Tlatelolco Regime and Nonproliferation in Latin America," *International Organization*, vol. 35, no. 1, Winter 1981, 103–34.

Reis Pereira, Guilherme, *Política espacial brasileira e a trajetória do INPE (1961–2007)*, Campinas, Brazil: Universidade Estadual de Campinas, 2008 [doctoral dissertation].

Reynolds, Glenn H. and Merges, Robert P., *Outer Space: Problems of Law and Policy*, 2nd edn., Boulder, CO: Westview Press, 1997.

Ross, Robert, *Negotiating Cooperation: The United States and China, 1969–1989*, Palo Alto, CA: Stanford University Press, 1995.

Ryzenko, Jakub, "Space in Poland—A Promising Future," *Space Policy*, vol. 19, no. 3, August 2003, 211–13.

Sadeh, Eligar (ed.), *Space Politics and Policy: An Evolutionary Perspective*, Dordrecht, Netherlands: Kluwer Academic Publishers, 2002.

Sagan, Carl, *Pale Blue Dot: A Vision of the Human Future in Space*, New York: Ballantine Books, 1994.

Sagan, Scott D., "Why Do States Build Nuclear Weapons? Three Models in Search of a Bomb," *International Security*, vol. 21, no. 3, Winter 1996–1997, 54–86.

Saich, Tony, "The Fourteenth Party Congress: A Programme for Authoritarian Rule," *The China Quarterly*, no. 132, December 1992, 1136–60.

Said, Mohamed Kadry, "Missile Proliferation in the Middle East: A Regional Perspective," *Disarmament Forum*, no. 2, 2001, 123–33.

Sánchez Peña, Miguel, "Sounding Rockets in Argentina," *Space Research in Argentina*, Buenos Aires: National Commission of Space Research, 1999.

Sánchez Peña, Miguel, "Scientific Experiences using Argentinean Sounding Rockets in Antarctica," *Acta Astronautica*, vol. 47, nos. 2–9, November 2000, 301–7.

Santoro, Daniel, *Operación Cóndor II: La historia secreta del misil que desmanteló Menem*, Buenos Aires: Ediciones Letra Buena, 1992.

Saunders, Phillip C., "China's Future in Space: Implications for U.S. Security," *Ad Astra: The Magazine for the National Space Society*, vol. 17, no. 1, Spring 2005, 21–2.

Scheetz, Thomas, "A Peace Dividend in South America? Defense Conversion in Argentina and Chile," in Baltagi, Badi H. and Sadka, Efraim (eds.), *The Peace Dividend*, Bingley, UK: Emerald Group Publishing Limited, 1996, 403–23.

Schofield, Clive and Storey, Ian, "Energy Security and Southeast Asia: The Impact on Maritime Boundary and Territorial Disputes," *Harvard Asia Quarterly*, vol. 9, no. 4, Fall 2005, 36–46.

Seiler, John, "South African Perspectives and Responses to External Pressures," *The Journal of Modern African Studies*, vol. 13, no. 3, 1975, 447–68.

Shah, Aqil, "Democracy on Hold in Pakistan," *Journal of Democracy*, vol. 13, no. 1, 2002, 69–76.

Shapir, Yiftah, "Iran's Efforts to Conquer Space," *Strategic Assessment*, vol. 8, no. 3, November 2005, available www.inss.org.il/publications.php?cat=21&incat=&read=160 (accessed 9 June 2010).

Sheehan, Michael, *The International Politics of Space*, London: Routledge, 2007.

Sheldon, John B., "A Really Hard Case: Iranian Space Ambitions and the Prospects for U.S. Engagement," *Astropolitics*, vol. 4, no. 2, Summer 2006: 229–51.

Sheppard, Ben and DeVore, Howard, *China's Aerospace and Defence Industry*, London: Jane's Information Group, December 2000.

Siddiqi, Asif Azam, "The Soviet Co-Orbital Anti-Satellite System: A Synopsis," *Journal of the British Interplanetary Society*, June 1997, 225–40.

Siddiqi, Asif A., *The Soviet Space Race with Apollo*, Gainesville, FL: University Press of Florida, 2003.

Soares de Lima, Maria Regina and Hirst, Monica, "Brazil as an Intermediate State and Regional Power: Action, Choice and Responsibilities," *International Affairs*, vol. 81, no. 1, 2006, 21–40.

Sobhani, Sohrab, *The Pragmatic Entente: Israeli–Iranian Relations, 1948–1988*, Westport, CT: Praeger Publishers, 1989.

Spector, Leonard S., *Nuclear Ambitions*, Boulder, CO: Westview Press, 1990.

Stumpf, Waldo, "South Africa: Nuclear Technology and Nonproliferation," *Security Dialogue*, no. 4, 1993, 497–8.

Sundararajan, Venkatesan, "Emerging Space Powers—A Comparative Study of National Policy and Economic Analysis for Asian Space Programs," paper presented at the 2006 conference of the American Institute of Aeronautics and Astronautics, San Jose, CA, 2006.

Talbott, Strobe, "Dealing with the Bomb in South Asia," *Foreign Affairs*, vol. 78, no. 2, 1999, 110–22.

Tarikhi, Parviz, "Iran's Space Programme: Riding High for Peace and Pride," *Space Policy*, vol. 25, no. 3, August 2009, 160–73.

Tellis, Ashley J., "Punching the U.S. Military's 'Soft Ribs': China's Anti-Satellite Weapon Test in Strategic Perspective," Carnegie Endowment for International Peace Policy Brief no. 51, June 2007.

Thompson, Joseph E., "Virtual Regime: A New Actor in the Geopolitical Arena," *PS: Political Science & Politics*, vol. 35, no. 3, 2002, 507–8.

Timmerman, Kenneth R., *The Death Lobby: How the West Armed Iraq*, Boston: Houghton-Mifflin, 1991.

Tiwari, Arun, *Wings of Fire: An Autobiography of APJ Abdul Kalam*, London: Sangam Books Ltd, 1st edn., 1999.

Turnball, Stephan, *Siege Weapons of the Far East: AD 960–1644*, vol. 2, Essex, UK: Osprey Publishing, 2002.

United States Central Intelligence Agency, "National Intelligence Estimate 11–1–67: The Soviet Space Program," Washington, DC, 2 March 1967.

Venter, Al J., *Iran's Nuclear Option: Tehran's Quest for the Atom Bomb*, Havertown, PA: Casemate, 2005.

Wang, Chunyuan, "China's Space Industry and its Strategy of International Cooperation," Center for International Security and Arms Control, Stanford University, July 1996, available http://iis-db.stanford.edu/pubs/10223/img-3261431–00012.pdf

Ward, Bob, *Dr. Space: The Life of Wernher Von Braun*, Annapolis, MD: Naval Institute Press, 2005.

Wehling, Fred, "Russian Nuclear and Missile Exports to Iran," *The Nonproliferation Review*, Winter 1999, 134–43.

Wheelon, Albert D., "Lifting the Veil on Corona," *Space Policy*, vol. 11, no. 4, November 1995, 249–60.

Wolter, Detlev, *Common Security in Outer Space and International Law*, New York, United Nations Publications, 2006.

Xu, Jianguo, "The Launching of Aussat," *Beijing Review*, 2 November 1992, 17–20.

Zaborsky, Victor, "Economics vs. Nonproliferation: US Launch Quota Policy toward Russia, Ukraine, and China," *The Nonproliferation Review*, vol. 7, no. 3, Autumn 2000, 152–61.

Zaborsky, Victor, "The Brazilian Export Control System," *The Nonproliferation Review*, vol. 10, no. 2, Summer 2003, 123–35.

Zahl, Harold A., *Electrons Away: Tales of a Government Secret*, New York: Vantage Point Press, 1968.

Zhang, Ming, "China's Changing Nuclear Posture: Reactions to the South Asian Nuclear Tests," Washington, DC: Carnegie Endowment for International Peace, 1998.

Zhao, Yun, "The 2002 Space Cooperation Protocol between China and Brazil: An Excellent Example of South–South Cooperation," *Space Policy*, vol. 21, no. 3, 2005, 213–19.

Zhu, Yilina and Xu, Fuxiang, "Status and Prospects of China's Space Programme," *Space Policy*, vol. 13, no. 1, February 1997, 69–75.

Index

Chinese space program: Academy of Space Technology 90; Aerospace Science and Technology Corporation 90; *bingshang* (soldiers in business) 89; cooperation agreements with other countries 91, 95; cooperation with Soviet Union 84; Divestiture Act of 1998 89; domestic political benefits 100; early missiles and nuclear tests 85; economic shift 87; Fifth Research Academy 84; first liquid fueled rocket 85; first indigenous satellite 89; first military satellite 97; first satellite 86; Four Modernizations 87; Group 581, 84; growing satellite industry 88; internationalism in space policy 87; launch services 90, 93, 95; *Long March* launcher 86, 100; lunar program 98; manned program 86, 89, 91, 92, 97; Mars probe 100; military benefits against US 96, 99; Plan 863, 89; Project 651, 86; Project 1059 84; reaction to *Sputnik* 83; satellite production 88, 97; socioeconomic benefits 95; space diplomacy 91; space plane 100; space policy vis à vis US 82; space station 4, 93

Churchill, Winston 6, 33, 37, 66
Colombia: Colombian Space Commission (CCE) 161; space policy white papers 162
Columbia shuttle 3
comprehensive national power 79
Cóndor missile: Argentina's competition with Brazil 112; development 149–50, 153; and Egypt 149, 166; and Iraq 142
Corona 52
cryogenic rocket engines 104, 106

Deng Xiaoping 87
developing countries: definition 12; China as developing country 12–13
Disaster Monitoring Constellation 169, 170

Eisenhower, Dwight D.: and International Geophysical Year 43–4; and *Sputnik* 47
Ecuador: Center for National Resource Extraction by Remote Sensing (CLIRSEN) 158; Ecuadorian Civilian Space Agency (EXA) 158–9; objective of space program 159
Egypt: early rocket development 166; first satellite 167; spy satellite 167
emerging space actors 29; classification scheme 78–9; first tier (criteria) 79–80; rationale for investment in space 74–5
European Space Agency: competition with China 90; ESA space policy 69; *Europa* 67; founding of 68

France: ballistic missiles development 67–8; *force de frappe* 67; *Gerboise Bleue* 67; space program 68

General Electric 34, 37
Great Wall Industry Corporation 88; implicated in missile proliferation 89

Indian space program: A.P.J. Abdul Kalam 101; Defense Research and Development Organization 104; first satellite 104; *Geosynchronous Satellite Launch Vehicle* 106; moon probe 106; nuclear program 104–5; plan to counter China 104; *Polar Satellite Launch Vehicle* 105–6; and security concerns 102–3; space plane 107
Indian Space Research Organization 104; budget 107; mission statement 108
Indonesia: space agency 182; satellites 182–3; launcher project 183
Inter-Islamic Network on Space Sciences and Technology (ISNET) 193
International Geophysical Year 42–4
international relations theory: international liberalism 25, 45; realism 17, 25, 30; Thucydides 37; Treaty of Westphalia 28
Iran: Aerospace Research Institute 132; cooperation with Israel 127; *Mesbah* satellite 128; Shah Reza Pahlavi 125; pre-revolutionary missile and space ambitions 125–6; Iran–Iraq War 127; *Omid* satellite 131; *Safir* launcher 130; *Shahab* missile series 128–9; *Sinah* satellite 130; socioeconomic applications of space program 132; *Zohreh* project 128
Iranian Space Agency 129; budget 132